家庭服务业从业人员岗位技能培训丛书

# 家政实用英语

## 编委会

主　　任：张丽丽　谢玲丽
副 主 任：赵建德　张宝霞

主　　编：徐　君　缪俊聪
主　　审：章　简

中国劳动社会保障出版社

**图书在版编目（CIP）数据**

家政实用英语/上海市家庭服务业行业协会组织编写. -- 北京：中国劳动社会保障出版社，2017

（家庭服务业从业人员岗位技能培训丛书）

ISBN 978－7－5167－2921－2

Ⅰ.①家… Ⅱ.①上… Ⅲ.①家政学-英语-岗位培训-教材 Ⅳ.①TS976

中国版本图书馆 CIP 数据核字（2017）第 308273 号

**中国劳动社会保障出版社出版发行**

（北京市惠新东街1号 邮政编码：100029）

\*

三河市华骏印务包装有限公司印刷装订 新华书店经销

787 毫米×1092 毫米 16 开本 8.5 印张 155 千字

2017 年 12 月第 1 版 2017 年 12 月第 1 次印刷

定价：28.00 元

读者服务部电话：（010）64929211/84209103/84626437

营销部电话：（010）84414641

出版社网址：http://www.class.com.cn

版权专有 侵权必究

如有印装差错，请与本社联系调换：（010）50948191

我社将与版权执法机关配合，大力打击盗印、销售和使用盗版图书活动，敬请广大读者协助举报，经查实将给予举报者奖励。

举报电话：（010）64954652

# 内 容 简 介

随着中国经济的发展，外籍人士在中国就业、定居的人数也日益增多，对家政服务员的能力提出了新的要求，简单的英语交流能力是家政服务员必备能力之一。本书的编写团队在涉外家政方面具有多年的培训经验，教材内容根据家政服务员的群体特点编写，易教学、易理解、易掌握。

本书涵盖了家政服务员在服务外籍家庭时的常用单词、词组和语句。本书共20课，设计了从自我介绍、招聘面试到涉外家政方方面面工作的20个场景。

每一课包括 Dialogue（情景对话）、New Words（新单词）、Useful Sentences（实用例句）、Grammar（语法）、Extension（扩展）和 Exercises（练习）。通过学习本书，涉外家政服务员可以基本掌握工作中的常用英语，进行简单的英语交流。

# 目 录

- Unit 1　Self Introduction & Daily Greetings ⋯⋯⋯⋯ 1
  第 1 课　自我介绍及日常问候 ⋯⋯⋯⋯⋯⋯⋯⋯⋯⋯ 1

- Unit 2　Recruitment & Interview ⋯⋯⋯⋯⋯⋯⋯⋯ 7
  第 2 课　招聘及面试常用语 ⋯⋯⋯⋯⋯⋯⋯⋯⋯⋯ 7

- Unit 3　Privacy & Confidentiality ⋯⋯⋯⋯⋯⋯⋯⋯ 13
  第 3 课　雇主及自身隐私保护用语 ⋯⋯⋯⋯⋯⋯⋯⋯ 13

- Unit 4　Community & Property Introduction ⋯⋯⋯⋯ 19
  第 4 课　社区物业介绍 ⋯⋯⋯⋯⋯⋯⋯⋯⋯⋯⋯⋯ 19

- Unit 5　Daily Cleaning ⋯⋯⋯⋯⋯⋯⋯⋯⋯⋯⋯⋯ 25
  第 5 课　日常保洁用语 ⋯⋯⋯⋯⋯⋯⋯⋯⋯⋯⋯⋯ 25

- Unit 6　Clothes, Laundry & Ironing ⋯⋯⋯⋯⋯⋯⋯ 31
  第 6 课　衣物及洗烫常用语 ⋯⋯⋯⋯⋯⋯⋯⋯⋯⋯ 31

- Unit 7　Cooking & Dinning ⋯⋯⋯⋯⋯⋯⋯⋯⋯⋯ 37
  第 7 课　烹饪及餐饮礼仪用语 ⋯⋯⋯⋯⋯⋯⋯⋯⋯ 37

- Unit 8　Housing Facilities & Home Appliances ⋯⋯⋯ 43

第 8 课　房屋设备及家用电器用语 ………………… 43

Unit 9　Child Care & Daily Education ………………… 49
第 9 课　孩子照护及日常教育常用语 ………………… 49

Unit 10　Pets, Animals & Plants ………………… 55
第 10 课　宠物及相关动植物英文名称 ………………… 55

Unit 11　Family Party & Social Activities ………………… 61
第 11 课　家庭聚会及社交活动用语 ………………… 61

Unit 12　Physiology & Common Diseases ………………… 67
第 12 课　生理构造及常见疾病名称 ………………… 67

Unit 13　Ask for Leave, Thanks & Apology ………………… 75
第 13 课　请假、致谢、致歉等用语 ………………… 75

Unit 14　Travel & Holidays ………………… 81
第 14 课　旅游度假常用语 ………………… 81

Unit 15　Working Hours & Salary ………………… 87

第 15 课　工作时间及薪资 ·················· 87

Unit 16　Budget & Expense Report ·············· 93
第 16 课　周期性预算及报账用语 ············ 93

Unit 17　First-Aid & Accident Prevention ········ 99
第 17 课　常用急救知识及意外防范用语 ········ 99

Unit 18　Dress Code Courtesy & Etiquette ········ 105
第 18 课　个人着装及礼貌礼仪用语 ············ 105

Unit 19　Religious Custom & Taboo ············· 111
第 19 课　宗教习俗及相关忌讳用语 ············ 111

Unit 20　Resignation & Dismissal ··············· 117
第 20 课　请辞用语 ······················· 117

附录　音标 ···························· 123

# Unit 1
## 第 1 课

自我介绍及日常问候
Self Introduction & Daily Greetings

家政实用英语

## Dialogue  情景对话

Linda：Good morning, sir.

琳达：早上好，先生。

Richard：Good morning.

理查德：早上好。

Linda：I am Linda, I want to work for you.

琳达：我叫琳达，我希望为你工作。

Richard：OK. Could you please introduce yourself?

理查德：好的，你能介绍一下你自己吗?

Linda：Yes. My name is Linda, I am from Shanghai, I am thirty-five years old. I have been working as a housekeeper for ten years.

琳达：可以，我的名字叫琳达，我来自上海，我35岁。我从事家政工作10年了。

Richard：Ten years! That is very long.

理查德：10年了！真是太长了。

Linda：I can clean very well, do the laundry well and cook.

琳达：我打扫很干净，洗衣服也很好，还会做饭。

Richard：You are a good housekeeper. Do you have any reference?

理查德：你真是个很好的家政服务员。你有介绍信吗?

Linda：Yes, I have. Here you are.

琳达：是的，我有介绍信。给你看一下。

Richard：Good. I will hire you, can you start tomorrow?

理查德：很好，我决定聘用你，你明天能开始上班吗?

Linda：Yes, I am happy to work for you.

琳达：当然，我很乐意为你工作。

Richard：See you tomorrow.

理查德：明天见。

Linda：See you.

琳达：明天见。

自我介绍及日常问候

## New Words 新单词

| 英语单词 | 音标 | 发音参考 | 中文翻译 |
|---|---|---|---|
| good | [ gʊd ] | 顾的 | 好 |
| morning | [ ˈmɔːnɪŋ ] | 末宁 | 早上,上午 |
| sir | [ sɜː ] | 色 | 先生,老师 |
| I | [ aɪ ] | 爱 | 我 |
| want | [ wɒnt ] | 忘特 | 需要,希望 |
| work | [ wɜːk ] | 我克 | 工作,事业 |
| for | [ fə; fɔː ] | 发 | 为,为了,因为 |
| you | [ juː ] | 油 | 你,你们 |
| OK | [ ˌəʊˈkeɪ ] | 欧可 | 好的,不错的 |
| could | [ kʊd ] | 苦的 | 能够 |
| please | [ pliːz ] | 怕立兹 | 请,使高兴 |
| introduce | [ ˌɪntrəˈdjuːs ] | 因尺丢思 | 介绍,提出 |
| yourself | [ jɔːˈself ] | 姚赛尔夫 | 你自己 |
| yes | [ jes ] | 叶思 | 是,是的 |
| my | [ maɪ ] | 麦 | 我的 |
| name | [ neɪm ] | 内母 | 名字 |
| am | [ em ] | 爱母 | 是,用于第一人称单数现在时 |
| from | [ frɒm ] | 福朗牡 | 从,由于 |
| thirty | [ ˈθɜːti ] | 色提 | 三十 |
| five | [ faɪv ] | 飞五 | 五,五个 |
| year | [ jɪə ] | 叶尔 | 年,年纪,年度 |
| old | [ əʊld ] | 欧尔得 | 老的,以前的 |
| as | [ æz ] | 爱自 | 作为 |
| housekeeper | [ ˈhaʊsˌkiːpə ] | 号思可以拍 | 管家,家政服务员 |
| ten | [ ten ] | 天 | 十,十个 |
| that | [ ðæt ] | 砸特 | 那,那个 |

**家政实用英语**

| 英语单词 | 音标 | 发音参考 | 中文翻译 |
|---|---|---|---|
| is | [ɪz] | 忆自 | 是 |
| very | [ˈveri] | 维瑞 | 很，非常，十足的 |
| long | [lɒŋ] | 狼 | 长的，遥远的 |
| clean | [kliːn] | 克林 | 打扫；干净的 |
| well | [wel] | 威尔 | 很好，良好地 |
| laundry | [ˈlɔːndri] | 狼坠 | 洗衣；洗衣房，洗好的衣服 |
| cook | [kʊk] | 库克 | 烹饪；厨师 |
| have | [hæv] | 哈五 | 有，让，拿，吃 |
| reference | [ˈrefrəns] | 来福润思 | 介绍信，参考，证明 |
| here | [hɪə] | 黑尔 | 这里，此时 |
| will | [wɪl] | 威尔 | 愿意；决心，医嘱 |
| hire | [ˈhaɪə] | 海尔 | 聘用，录用，雇用 |
| can | [kæn] | 肯 | 可以，能够，可能 |
| start | [stɑːt] | 思达特 | 开始，动身；起点 |
| tomorrow | [təˈmɒrəʊ] | 特磨柔 | 明天，未来 |
| happy | [ˈhæpi] | 海皮 | 快乐的，幸福的 |
| see | [siː] | 西 | 看，看见，理解 |

## Useful Sentences 实用例句

My name is Linda. 我的名字叫琳达。
I am from Shanghai. 我来自上海。

## Grammar 语法

**人称代词**：代词用来代替名词，在句子中起名词的作用。人称代词是指直接指代人或者事物的代词。人称代词一共分为三种：第一人称（我、我们），第二人称（你、你们），第三人称（他、她、它，他们、她们、它们）。

| 英语单词 | 音标 | 发音参考 | 中文翻译 |
|---|---|---|---|
| I | [aɪ] | 爱 | 我 |

自我介绍及日常问候

| 英语单词 | 音标 | 发音参考 | 中文翻译 |
|---|---|---|---|
| we | [wiː] | 威 | 我们 |
| you | [ju] | 又 | 你，你们 |
| he | [hiː] | 黑 | 他 |
| she | [ʃiː] | 戏 | 她 |
| it | [ɪt] | 一特 | 它 |
| they | [ðeɪ] | 贼 | 他们，她们，它们 |

例句：

I am from Anhui. 我来自安徽。

You are a great housekeeper. 你是个很棒的家政服务员。

I am happy to work for you. 我很乐意为你工作。

It is good. 这很好。

We will hire you. 我们决定聘用你。

He is from Shanghai. 他来自上海。

## Extension 扩展

English Letters 英语字母

| A a | B b | C c | D d | E e | F f | G g |
|---|---|---|---|---|---|---|
| H h | I i | J j | K k | L l | M m | N n |
| O o | P p | Q q | R r | S s | T t | |
| U u | V v | W w | X x | Y y | Z z | |

## Exercises 练习

Dialogue Practice　对话练习

1. What's your name?

2. Are you from Anhui?

3. How old are you?

4. Can you cook?

5. Can I have a look at your reference?

**Possible Answers** 参考答案

1. What's your name?   — My name is ×××.
2. Are you from Anhui?   — Yes, I am from Anhui.
3. How old are you?   — I am ×× years old.
4. Can you cook?   — Yes, I can.
5. Can I have a look at your reference?   — Yes, here you are.

# Unit 2
# 第 2 课

招聘及面试常用语
Recruitment & Interview

**Dialogue  情景对话**

Lisa: Good afternoon, madam. I am here to apply for the housekeeper job.

莉萨: 下午好, 夫人。我来申请家政服务员的工作。

Emily: Welcome. Could you please introduce yourself?

艾米莉: 欢迎欢迎。你能自我介绍一下吗?

Lisa: Sure. My name is Lisa. I am from Anhui. I am 35 years old. I have been working as a housekeeper for five years.

莉萨: 好的, 我名字叫莉萨, 来自安徽, 现在35岁。我从事家政工作5年了。

Emily: That sounds good. What can you do?

艾米莉: 听着不错。你会做什么?

Lisa: I can clean, iron, do laundry, go shopping and cook both Chinese and Western food.

莉萨: 我会保洁、熨烫、洗衣服、买东西、做中餐和西餐。

Emily: Great. Can you cook Xiaolongbao?

艾米莉: 太棒了。你会做小笼包吗?

Lisa: Yes, I am very good at cooking.

莉萨: 是的, 我很擅长烹饪。

Emily: Good to know this. Can you also help with child care?

艾米莉: 很高兴知道这个。你可以帮我照顾孩子吗?

Lisa: Yes, I like children, I know how to play with them and discipline them.

莉萨: 是的, 我喜欢孩子。我知道怎么陪孩子玩, 也知道怎么管教孩子。

Emily: OK. I need 5 days per week from 8am to 6pm. Is it OK for you?

艾米莉: 好的, 我需要一周工作5天, 上午8点到下午6点。这样可以吗?

Lisa: Yes, the schedule is great. I was paid RMB 7000 per month before, is it OK for you?

莉萨: 好的, 工作时间非常好。我以前的工作, 月工资7000元, 您看可以吗?

Emily: OK. Can you start tomorrow?

艾米莉: 好的, 你明天能开始吗?

Lisa: Yes, no problem. See you tomorrow at 8am.

莉萨: 好的, 没问题。明天上午8点见。

Emily：See you tomorrow.
艾米莉：明天见。

### New Words 新单词

| 英语单词 | 音标 | 发音参考 | 中文翻译 |
| --- | --- | --- | --- |
| recruitment | [rɪˈkruːtmənt] | 日库入特门特 | 招聘，补充 |
| interview | [ˈɪntəvjuː] | 因特吴由 | 面试，采访 |
| afternoon | [ɑːftəˈnuːn] | 阿夫特浓 | 下午，午后 |
| madam | [ˈmædəm] | 麦登母 | 夫人，女士 |
| apply | [əˈplaɪ] | 呃彭来 | 申请，涂，敷 |
| job | [dʒɒb] | 脚步 | 工作，职业 |
| welcome | [ˈwelkəm] | 威尔康姆 | 欢迎 |
| sure | [ʃɔː] | 肖尔 | 确信的，可靠的 |
| sound | [saʊnd] | 桑德 | 听起来；声音，语音 |
| what | [wɒt] | 瓦特 | 什么，多少，多么 |
| iron | [ˈaɪən] | 爱恩 | 烫衣服；熨斗 |
| go shopping | [gəʊ] [ˈʃɒpɪn] | 够小拼 | 购物，买东西 |
| both | [bəʊθ] | 不欧思 | 两个的，双方的 |
| Chinese | [ˌtʃaɪˈniːz] | 钗腻思 | 中国人，中文，汉语；中国的 |
| Western | [ˈwest(ə)n] | 威思特恩 | 西方的；西方人，西部片 |
| food | [fuːd] | 富得 | 食物，养分 |
| great | [greɪt] | 格瑞特 | 太棒了，伟大的；大师，伟人 |
| know | [nəʊ] | 弄 | 了解，熟悉，知道 |
| this | [ðɪs] | 姐思 | 这，这个，这么 |
| also | [ˈɔːlsəʊ] | 奥尔搜 | 另外，并且，也 |
| help | [help] | 黑尔普 | 帮助，促进，补救 |
| with | [wɪð] | 威姿 | 和……在一起，随着 |
| child | [tʃaɪld] | 巧儿得 | 儿童，小孩，子孙 |

## 家政实用英语

| 英语单词 | 音标 | 发音参考 | 中文翻译 |
|---|---|---|---|
| care | [keə] | 开尔 | 关怀，照料，在意，顾虑 |
| children | ['tʃɪldrən] | 巧儿群 | 孩子们（child 的复数） |
| how | [haʊ] | 号 | 如何，多么 |
| play | [pleɪ] | 泼雷 | 玩耍；游戏，比赛 |
| them | [ðem] | 增母 | 他们，她们 |
| discipline | ['dɪsɪplɪn] | 地西普林 | 管教，训练；纪律 |
| need | [niːd] | 拟得 | 需求，需要 |
| day | [deɪ] | 得 | 一天，白天 |
| per | [pɜː] | 坡 | 每，经，按照 |
| week | [wiːk] | 威克 | 星期，周 |
| am | [ˌeɪˈem] | 诶爱慕 | 上午，午前 |
| pm | [ˌpiːˈɛm] | 皮爱母 | 下午 |
| schedule | ['skɛdʒəl] | 思凯久 | 时间表，计划表，安排 |
| pay | [peɪ] | 配 | 支付，偿还；工资，薪水 |
| month | [mʌnθ] | 忙思 | 月，月份 |
| before | [bɪˈfɔː] | 比佛 | 在……之前，先于 |
| problem | ['prɒbləm] | 普罗不伦 | 问题，难题，麻烦 |

### Useful Sentences  实用例句

Could you please introduce yourself?  你能自我介绍一下吗？
I can clean, iron, do laundry, go shopping and cook both Chinese and Western food.
我会保洁、熨烫、洗衣服、买东西、做中餐和西餐。

### Grammar  语法

**情态动词**：本身有一定的意义，表示语气的单词，一般表示看法或主观判断。情态动词虽然数量不多，但用途广泛，主要的情态动词有 can, could, may, might, must, need 等。

can：表示能力，可能性，有时会，或者允许。

could：作为 can 的过去式，表示过去的情况；或者表示能力，可能性；或者婉转地提出请求、想法、建议等。

例句：

Can you come tomorrow? 你明天能来吗？

You can go now. 你可以走了。

Could you please introduce yourself? 你能自我介绍一下吗？

Can you start tomorrow? 你可以明天就开始吗？

 Extension 扩展

反身代词：反身代词是一种表示反射或强调的代词，如我自己、你自己、他自己、我们自己等。反身代词由人称代词宾格形式，词尾加 self（单数）或 selves（复数）组成，一般译为本人、本身、亲自等。反身代词主要有 myself, yourself, himself, herself, itself, ourselves, yourselves, themselves。

 Exercises 练习

Dialogue Practice  对话练习

1. Could you please introduce yourself?

2. What can you do?

3. Can you discipline children?

4. Is 10 hours per day OK for you?

5. Can you start today?

 Possible Answers 参考答案

1. Could you please introduce yourself?    — Yes, my name is ×××, I am from ×××, I am ×× years old.

2. What can you do?    — I can clean, cook and take care of baby.

3. Can you discipline children?    — Yes, I can.

4. Is 10 hours per day OK for you?    — Yes, it's OK.

5. Can you start today?    — Yes, I can start today.

Unit 3
第 3 课

雇主及自身隐私保护用语
Privacy & Confidentiality

**Dialogue 情景对话**

Cindy: Since we have finished the work schedule and welfare part, shall we move towards the obligation part?

辛迪：既然我们谈好了工作时间、福利待遇，我们来谈谈义务？

Lisa /Emily: OK. Let's continue.

莉萨/艾米莉：好的，我们继续。

Cindy: As an employee, you are not supposed to tell anybody about the employer's personal and family information. If there is any disclosure, you may face lawsuit and claim of compensation.

辛迪：雇员不应当对任何人谈起雇主的个人和家庭信息。如果涉及信息泄露，会导致诉讼和赔偿。

Lisa: Yes, I promise to keep confidentiality of privacy, not to tell anybody of anything related to this family, including my family members.

莉萨：是的，我保证我会保证隐私安全，不告诉任何人雇主家庭的任何信息，包括我的家人。

Cindy: And please do not bring anyone home without the employer's permission.

辛迪：没有雇主同意，不得带任何人到雇主家。

Lisa: Definitely, I won't bring anyone without Emily's prior consent.

莉萨：肯定的，艾米莉不同意，我不会带任何人来的。

Emily: Yes, but I would allow you to take your kids along sometimes, but I need to know in advance.

艾米莉：是的，我同意你可以带自己的孩子来，但是我要提前知道。

Cindy: OK. As to the employer's obligations, the employer shall always pay salary on time, with no delay. If your salary is delayed for more than one week, you will be paid extra compensations. And the employer shall not abuse the employee.

辛迪：好的，现在来谈谈雇主的责任和义务。雇主应当准时支付工资，不能推迟，如果推迟支付超过一周则需要支付额外的赔偿，并且不得虐待雇员。

Emily: Definitely. I always treat ayi (housekeeper) like my family.

艾米莉：当然了，我一直待阿姨（家政服务员）像家人一样。

Cindy: That's good. Do you two have anything else to add in the contract?

雇主及自身隐私保护用语

辛迪：很好，你们双方还有什么要加到合同里面的吗？
Lisa/Emily：No, thank you!
莉萨/艾米莉：没有了，谢谢！
Cindy：OK. We can sign now. Hope you two are happy with each other.
辛迪：好的，我们可以签字了。希望你们互相都满意。
Lisa/Emily：Definitely. Thank you, Cindy.
莉萨/艾米莉：当然了，谢谢你，辛迪。

## New Words 新单词

| 英语单词 | 音标 | 发音参考 | 中文翻译 |
| --- | --- | --- | --- |
| privacy | ['prɪvəsi] | 普瑞吴斯 | 隐私，秘密 |
| confidentiality | [ˌkɒnfɪdenʃi'ælɪti] | 康飞丹迅艾立特 | 机密 |
| since | [sɪns] | 心思 | 因为，由于，自从 |
| welfare | ['welfeə] | 威尔费尔 | 福利，幸福 |
| move | [muːv] | 木五 | 移动，搬家 |
| forward | ['fɔːwəd] | 佛我得 | 向前的，早的 |
| obligation | [ɒblɪ'geɪʃn] | 奥布立给迅 | 义务，职责 |
| employee | [ɪm'plɔɪ-iː] | 英宙普落叶 | 雇员，从业人员 |
| suppose | [sə'pəʊz] | 色坡兹 | 假设，认为 |
| personal | ['pɜːsənəl] | 普色诺 | 个人的，身体的 |
| disclosure | [dɪs'kləʊʒə] | 迪斯克楼许 | 披露，揭发 |
| face | [feɪs] | 菲思 | 面对，面临；脸 |
| claim | [kleɪm] | 克雷木 | 要求，声称，认领 |
| compensation | [kɒmpen'seɪʃn] | 康盘赛迅 | 补偿，报酬 |
| promise | ['prɒmɪs] | 普罗密斯 | 许诺，允许 |
| related | [rɪ'leɪtɪd] | 瑞雷提德 | 有关系的，有关联的 |
| without | [wɪ'ðaʊt] | 威造特 | 没有，超过 |
| permission | [pə'mɪʃn] | 普秘迅 | 允许，许可 |
| prior | ['praɪə] | 普瑞尔 | 优先的，在前面的 |
| consent | [kən'sent] | 肯森特 | 同意，赞成 |
| allow | [ə'laʊ] | 阿劳 | 允许，认可 |

家政实用英语

| 英语单词 | 音标 | 发音参考 | 中文翻译 |
|---|---|---|---|
| along | [əˈlɒŋ] | 阿郎 | 一起，向前，来到 |
| advance | [ədˈvɑːns] | 阿德旺斯 | 发展，前进；先进的 |
| in advance | [ɪn] [ədˈvɑːns] | 音阿德旺斯 | 提前 |
| abuse | [əˈbjuːz] | 阿比斯 | 虐待，滥用 |
| treat | [triːt] | 特力特 | 对待，治疗 |
| add | [æd] | 爱德 | 增加，添加 |
| contract | [ˈkɒntrækt] | 康崔克特 | 合同 |
| sign | [saɪn] | 赛恩 | 迹象，符号 |

## Useful Sentences 实用例句

Do not bring anyone home without employer's permission.
没有雇主同意，不得带任何人到雇主家。

Now come to the employer's obligations.
现在来谈谈雇主的责任和义务。

Hope you two are happy with each other.
希望你们互相都满意。

## Grammar 语法

引导原因状语从句的连词有 because，as，since 等，用来解释某件事发生的原因。

> 例句：
> Since we have finished the schedule and welfare part, shall we move towards the obligation part?
> 既然我们谈好了工作时间、福利待遇，我们来谈谈义务？
> Because everything is very expensive in big cities, we have to work hard to make more money.
> 因为大城市什么东西都很贵，我们需要努力工作赚更多的钱。
> As I am a working mother, I need to work harder to balance my work and life.
> 因为我是有工作的母亲，我需要更努力来平衡工作和生活。

### Extension 扩展

**工资的形式**

salary：工资，指工作所得的收入，一般是固定的收入，按照月或者年来计算。

wage：一般指体力劳动者按钟点、天数、周，或者计件所得收入，常用复数形式。

pay：指收入，不论工作性质，只要是劳动报酬，包含 salary 和 wage。

income：收入，收益，通常不针对一份工作的收入，而是整年的所有收入和收益所得。

例句：

He is a carpenter, he gets good wages every week. 他是个木匠，他每周的工资比较高。

She is making a good salary in bank. 她在银行工作工资很高。

How much does your boss pay you? 你老板付你多少工资？

His annual income is very high, including salary, rental and investments. 他的年收入很高，包括工资、房租收入和投资。

### Exercises 练习

Dialogue Practice  对话练习

1. When you have finished your work, you can go home.
2. Would you mind keeping the secret for me?
3. Would you mind if I pay salary on 7th of each month?
4. Can I ask for leave for 1 week? My father is sick.
5. It's time to sign the contract, are you OK with all terms?

### Possible Answers 参考答案

1. When you have finished your work, you can go home.　　— Thank you.
2. Would you mind keeping the secret for me?　　— OK.
3. Would you mind if I pay salary on 7th of each month?　　— No problem.
4. Can I ask for leave for 1 week? My father is sick.　　— OK.
5. It's time to sign the contract, are you OK with all terms?　　— No problem at all.

# Unit 4
# 第 4 课

社区物业介绍
Community & Property Introduction

家政实用英语

**Dialogue　情景对话**

Lisa： Good morning, Emily.

莉萨： 早上好，艾米莉。

Emily： Good morning. Welcome to our home.

艾米莉： 早上好。欢迎来我们家。

Lisa： You have a very nice house and beautiful garden.

莉萨： 你家的房子好漂亮，花园很美丽。

Emily： Thank you! I would like to show you around, and then register at the management office.

艾米莉： 谢谢你！我带你看一下我们家的房子，然后去物业登记一下。

Lisa： OK. Let's go.

莉萨： 好的，走吧。

Emily： Hi, John, I would like to introduce you my housekeeper Lisa. I also call her ayi.

艾米莉： 嗨，约翰，我来介绍一下我们家的家政服务员莉萨。我也叫她阿姨。

John： Hi Emily. Nice to meet you, Lisa.

约翰： 嗨，艾米莉。很高兴见到你，莉萨。

Lisa： Nice to meet you too, John. I am new here, is there anything I shall pay attention to?

莉萨： 我也很高兴见到你，约翰。我是新来的，有些什么要注意的吗？

John： Yes. First, we need your ID and photo for a pass. You must show the guard your pass every time. You can not lend your pass to anyone else.

约翰： 是的，首先我们需要你的身份证和照片办理出入证。每次进小区你都要给保安看一下你的出入证。出入证不能借给任何人。

Lisa： OK. Anything else?

莉萨： 好的，还有其他要注意的吗？

John： Second, club is open everyday from 9am to 9pm, if kids go to the swimming pool, you should accompany them. Third, if you need any support, you are welcome to call us anytime. Here is the number.

约翰： 其次，会所每天上午9点到晚上9点开放，如果孩子们去游泳池玩，你都要陪着。再次，如果你需要任何帮助，欢迎随时联系我们。这是电话号码。

Emily: Oh, yes, my kids love swimming and the play ground. When I am not at home, you can go with them.

艾米莉：噢，是的，我的孩子们都喜欢游泳和游乐场。我不在的时候，你可以带他们来玩。

Lisa: OK. And thank you, John.

莉萨：好的。谢谢你，约翰。

## New Words 新单词

| 英语单词 | 音标 | 发音参考 | 中文翻译 |
| --- | --- | --- | --- |
| community | [kəˈmjuːnəti] | 科木由那提 | 社区，团体 |
| property | [ˈprɒpəti] | 普若普提 | 财产，所有权 |
| introduction | [ˌɪntrəˈdʌkʃən] | 音出达克迅 | 介绍，入门 |
| our | [aʊə] | 奥尔 | 我们的 |
| home | [həʊm] | 厚木 | 家，住宅；家庭的 |
| nice | [naɪs] | 耐思 | 美好的，精密的，和蔼的 |
| beautiful | [ˈbjuːtɪful] | 不有特福 | 美丽的，出色的，迷人的 |
| garden | [ˈɡɑːd(ə)n] | 家登 | 花园，菜园；从事园艺 |
| thank | [θæŋk] | 三克 | 感谢，谢谢 |
| would | [wʊd] | 伍德 | 将，将要，愿意 |
| like | [laɪk] | 莱克 | 喜欢，愿意；爱好 |
| show | [ʃəʊ] | 秀 | 展示，说明 |
| around | [əˈraʊnd] | 阿让的 | 在附近，大约，到处 |
| then | [ðen] | 蚕 | 然后，于是，此外 |
| register | [ˈredʒɪstə] | 瑞吉思特 | 登记，注册，挂号 |
| management | [ˈmænɪdʒm(ə)nt] | 曼妮奇门特 | 管理；管理人员，管理部门 |
| office | [ˈɒfɪs] | 奥菲思 | 办公室，政府机关，营业处 |
| let | [let] | 莱特 | 让，允许，出租 |

## 家政实用英语

| 英语单词 | 音标 | 发音参考 | 中文翻译 |
|---|---|---|---|
| meet | [miːt] | 米特 | 遇见，满足，对付 |
| too | [tuː] | 兔 | 也，太，很，非常 |
| new | [njuː] | 拗 | 新的，新鲜的 |
| there | [ðeə] | 再尔 | 在那里，在那边 |
| anything | [ˈenɪθɪŋ] | 爱宁新 | 任何事情 |
| shall | [ʃæl] | 笑 | 应该，会 |
| attention | [əˈtenʃ(ə)n] | 爱谈讯 | 注意力，关心 |
| first | [fɜːst] | 发思特 | 首先，第一，优先 |
| we | [wiː] | 唯一 | 我们 |
| your | [jɔː] | 要 | 你的，你们的 |
| ID | [aɪˈdiː] | 爱弟 | 身份证 |
| photo | [ˈfəutəu] | 否偷 | 照片 |
| pass | [pɑːs] | 帕斯 | 通行证 |
| must | [mʌst] | 马斯特 | 必须，应当 |
| guard | [gɑːd] | 加德 | 保安，守卫 |
| every | [ˈevrɪ] | 爱维瑞 | 每一个，每个 |
| time | [taɪm] | 太母 | 次；时（间） |
| lend | [lend] | 兰德 | 把……借给 |
| anyone | [ˈenɪwʌn] | 爱宁晚 | 任何人，任何一个 |
| else | [els] | 艾尔思 | 另外，其他 |
| second | [ˈsek(ə)nd] | 三更德 | 第二，其次；秒 |
| club | [klʌb] | 克拉卜 | 俱乐部，社团，夜总会 |
| open | [ˈəup(ə)n] | 欧喷 | 打开；公开的，空旷的 |
| everyday | [ˈevrɪdeɪ] | 爱维瑞得 | 每天；日常的 |
| if | [ɪf] | 一夫 | 如果，假设，设想 |
| kid | [kɪd] | 开德 | 小孩 |
| swimming pool | [ˈswɪmɪŋ puːl] | 思威敏 普尔 | 游泳池 |
| should | [ʃud] | 晓得 | 应该，可能，将要 |
| accompany | [əˈkʌmpəni] | 阿康普你 | 陪伴，伴随 |

| 英语单词 | 音标 | 发音参考 | 中文翻译 |
|---|---|---|---|
| third | [θɜːd] | 涩的 | 第三，三分之一 |
| call | [kɔːl] | 考尔 | 呼叫，拜访，召集 |
| us | [ʌs] | 阿思 | 我们 |
| number | [ˈnʌmbə] | 囊不 | 号码，数量，数字 |
| love | [lʌv] | 拉夫 | 爱；恋爱，亲爱的 |
| ground | [ɡraʊnd] | 格让德 | 地面，土地，范围 |
| when | [wen] | 万 | 当……时，什么时候 |
| at | [æt] | 爱特 | 在，向 |

### Useful Sentences 实用例句

Welcome to our home. 欢迎来我们家。

I would like to show you around. 我带你看一下我们家的房子。

### Grammar 语法

**助动词1**：协助主要动词构成谓语的词叫助动词，也叫辅助动词。助动词用来构成时态和语态。助动词本身没有意义，不可单独作谓语。最常用的助动词有 be(am, is, are)、have, has, do, would, should 等。

> 例句：
> 
> I am new here. 我是新来的。
> 
> Is there anything I shall pay attention to? 有什么我需要注意的吗？
> 
> When I am not at home, you can go with them. 我不在家的时候，你可以和他们一起去。
> 
> I would like to show you around. 我带你转一圈看一下。
> 
> If kids go to the swimming pool, you should accompany them. 如果孩子们去游泳池，你要陪着他们。

时间和序列的用法

1. when 当……的时候

When I am not at home, you can come with them.

2. then 然后

I would like to show you around, then register at the management office.

3. 序列 First,.... Second,.... Third,....

First, we need your ID and photo for a pass card. Second, club is open every day from 9am to 9pm. Third, if you need any support, you are welcome to call us anytime.

### Exercises 练习

Dialogue Practice  对话练习

1. Hi, Lisa, welcome to our home.
2. I would like to show you around.
3. Nice to meet you.
4. Could you please show the pass to our guard every time?
5. When I am not at home, please go pick up David from school.

### Possible Answers 参考答案

1. Hi, Lisa, welcome to our home.       — Thank you.
2. I would like to show you around.     — OK, let's go.
3. Nice to meet you.       — Nice to meet you, too.
4. Could you please show the pass to our guard every time?       — OK, I will.
5. When I am not at home, please go pick up David from school.       — OK, I will.

# Unit 5
# 第 5 课

日常保洁用语
Daily Cleaning

### Dialogue 情景对话

Lisa: Hi Emily, good morning.

莉萨：嗨，艾米莉，早上好。

Emily: Good morning, Lisa. For the first day, I would like to tell you about our cleaning manner, is it OK for you?

艾米莉：早上好，莉萨。今天第一天，我跟你说一下我们家的保洁习惯，可以吗？

Lisa: Of course, I will respect your way.

莉萨：当然可以，我会尊重您家的方式。

Emily: Good. First of all, you should wash your hands whenever you come from outside, or after you cleaned the toilet, or before cooking.

艾米莉：好的。首先，不管什么时候从外面进来，或者打扫好马桶，或者做饭之前，都要先洗手。

Lisa: Yes, I always do the same.

莉萨：好的，我一直都是这么做的。

Emily: Good. Second, I need you to clean kitchen first, then dust, vacuum, and clean bathroom.

艾米莉：好的。其次，你先打扫厨房，然后擦灰、吸尘、打扫卫生间。

Lisa: Sure, I know. Start from the kitchen as always.

莉萨：当然，我知道。总是先打扫厨房。

Emily: You should always leave the mopping till the end, including mopping the balcony floor. And cleansers need to be used for proper cleaning purpose.

艾米莉：拖地应当总是放到最后做，包括拖阳台。为了打扫得干净，都要使用清洁剂。

Lisa: Sure. What about ironing?

莉萨：好的。那熨烫呢？

Emily: Please start ironing after all cleaning work is done.

艾米莉：请你保洁好了之后再熨烫。

Lisa: That's OK. May I have a pair of gloves?

莉萨：好的。我可以配副手套吗？

Emily: Of course. I have some new gloves. You can take one.

艾米莉：当然，我有一些新的手套，你可以拿一副。

Lisa: Thank you. Shall I start now?
莉萨：谢谢。我现在可以开始了吗？
Emily: Yes, go ahead.
艾米莉：好的，去忙吧。

## New Words 新单词

| 英语单词 | 音标 | 发音参考 | 中文翻译 |
| --- | --- | --- | --- |
| daily | [ˈdeɪli] | 得丽 | 日常的，每日的 |
| cleaning | [ˈkliːnɪŋ] | 克林宁 | 清洗，清除，去污 |
| tell | [tel] | 胎儿 | 告诉，说 |
| about | [əˈbaʊt] | 阿宝特 | 关于，大约 |
| manner | [ˈmænə] | 曼娜 | 方式，习惯，风俗 |
| of course | [ɒv] [kɔːs] | 欧无考斯 | 当然 |
| respect | [rɪˈspekt] | 瑞思贝克特 | 尊重，尊敬 |
| way | [weɪ] | 卫 | 方法，方式，道路，习惯 |
| wash | [wɒʃ] | 我洗 | 洗涤，冲刷 |
| hand | [hænd] | 汗的 | 手，手艺 |
| whenever | [wenˈevə] | 万爱五 | 每当，无论何时，随便什么时候 |
| come | [kʌm] | 康姆 | 来，开始，出现 |
| outside | [aʊtˈsaɪd] | 奥特赛德 | 外面的，外部的 |
| after | [ˈɑːftə] | 阿福特 | 在……之后，后来 |
| toilet | [ˈtɔɪlɪt] | 托儿利特 | 厕所，盥洗室 |
| always | [ˈɔːlweɪz] | 奥威姿 | 总是，永远 |
| do | [duː] | 杜 | 干，做，要求，的确 |
| same | [seɪm] | 赛母 | 相同的，同一的 |
| kitchen | [ˈkɪtʃɪn] | 开群 | 厨房，炊具 |
| dust | [dʌst] | 达思特 | 灰尘；掸灰 |
| vacuum | [ˈvækjuəm] | 万扣母 | 吸尘；真空 |

家政实用英语

| 英语单词 | 音标 | 发音参考 | 中文翻译 |
|---|---|---|---|
| bathroom | [ˈbɑːruːm] | 巴斯容 | 卫生间，厕所 |
| leave | [liːv] | 里无 | 留下，离开 |
| mopping | [ˈmɒpɪŋ] | 帽平 | 拖地 |
| including | [ɪnˈkluːdɪŋ] | 英库卢定 | 包含，包括 |
| balcony | [ˈbælkəni] | 拜尔科倪 | 阳台，包厢 |
| cleanser | [ˈklenzə] | 克兰泽 | 清洁剂，清洁用品 |
| use | [juːz] | 油字 | 使用，利用 |
| proper | [ˈprɒpə] | 普若普 | 适当的，特有的 |
| purpose | [ˈpɜːpəs] | 帕普思 | 目的，用途 |
| all | [ɔːl] | 哦尔 | 全部的，全部 |
| done | [dʌn] | 当 | 已经完成的，做好了的 |
| pair | [peə] | 派儿 | 一对，一双 |
| gloves | [glʌvz] | 哥拉五子 | 手套（复数形式） |
| some | [sʌm] | 三母 | 一些，大约 |
| now | [naʊ] | 闹 | 现在，立刻 |

## Useful Sentences 实用例句

You should always leave the mopping till the end, including mopping the balcony floor.
拖地应当总是放到最后，包括拖阳台。

May I have a pair of gloves? 我可以配副手套吗？

## Grammar 语法

**助动词 2**：除了常用的助动词外，还有一些特殊的助动词，如 shall，should，它们虽然是助动词，但有时候有情态动词的作用，是介于助动词和情态动词之间的词汇，作用类似于 can，may，must。

shall：表示将要、命令、许诺等。主要用于第一人称，用来命令，征询意见。常用于正式文字中表示允诺、命令或者法令。

should：表示有义务的，有必要的，应当的，或者表示惊讶。也可在某些条件中代替 shall。

例句：

Shall I come tomorrow? 我明天要来吗？

Thomas shall be back anytime. 托马斯会随时回来。

Payment shall be made by cash. 请用现金付款。

You should wash your hands whenever you come from outside. 你从外面回来的时候都要洗手。

We should say thank you to you. 我们要对你说感谢。

It's 5pm now, Thomas should be at home now. 现在5点了，托马斯应该在家了。

### Extension 扩展

first of all（first），…；second，…；third，… 用来表示顺序，首先，……；其次，……；再次，……

First of all, you should wash your hands; second, I need you to clean kitchen; third, mop the floor.

### Exercises 练习

Dialogue Practice  对话练习

1. Is it OK for you to come this weekend.

2. Could you please clean the kitchen for me?

3. You should always use cleaning products.

4. May I have a pair of gloves?

5. Shall I start now?

### Possible Answers 参考答案

1. Is it OK for you to come this weekend?     — Yes.

2. Could you please clean the kitchen for me?     — OK, no problem.

3. You should always use cleaning pruducts.     — Yes, I will.

4. May I have a pair of gloves?     — Of course.

5. Shall I start now?     — Yes.

# Unit 6
# 第 6 课

衣物及洗烫常用语
Clothes, Laundry & Ironing

### Dialogue 情景对话

Emily: Hi Lisa, summer is coming, we shall start to wash our winter clothes and put them away.

艾米莉：嗨，莉萨，夏天快到了，冬天的衣服要开始洗干净、收纳好了。

Lisa: Yes, I am planning to do this after the rain season, or clothes will go moldy.

莉萨：是的，我打算在雨季以后再整理收纳。雨季衣服会发霉的。

Emily: Oh, really? Is the mould that bad in Shanghai?

艾米莉：哦，真的吗？上海那么容易发霉吗？

Lisa: Yes, it rains for 1-2 months before summer. If you don't have air conditioner or dehumidifier on all the time, everything goes mouldy.

莉萨：是的，夏天以前雨季要持续1~2个月。如果不一直开着空调或者除湿器，所有的东西都会发霉。

Emily: OK. We can delay this. But I also have some suggestions for ironing.

艾米莉：好的，那我们晚点收纳。但是我有一些关于熨烫的建议。

Lisa: I am happy to take your advice.

莉萨：我很乐意接受你的意见。

Emily: For my overcoat and suit, I usually send out for dry cleaning. But for jacket, shirt, pants and jeans, they are usually ironed at home.

艾米莉：我的大衣、西服等一般都是送出去干洗的。但是短外套、衬衣、裤子、牛仔裤都是在家熨烫的。

Lisa: Yes, I also iron bedding, so they look good like hotels'.

莉萨：是的，我也熨烫床单，这样看着就像酒店一样干净整洁了。

Emily: Thank you, but I need my husband's shirts to be ironed better, especially the collar and cuff. I want him to look smart.

艾米莉：谢谢你，我希望我老公的衬衣烫得更挺一些，尤其是领子和袖口。我希望他看着更精神。

Lisa: OK. I will pay more attention.

莉萨：好的，我以后多注意一点。

Emily: But you do clean very well, just look like five star hotels.

艾米莉：但是你打扫得非常好，就像五星级酒店一样干净。

Lisa: Thank you! I will keep my standard.

莉萨：谢谢！我一定保持我的保洁水准。

### New Words 新单词

| 英语单词 | 音标 | 发音参考 | 中文翻译 |
| --- | --- | --- | --- |
| clothes | [kləʊðz] | 克楼姿 | 衣服（复数） |
| summer | [ˈsʌmə] | 沙漠 | 夏季；夏天的 |
| winter | [ˈwɪntə] | 温特 | 冬季；冬天的 |
| put | [pʊt] | 普特 | 放，安置 |
| away | [əˈweɪ] | 阿魏 | 放到，放进（安全或封闭的地方） |
| plan | [plæn] | 普兰 | 计划，规划，打算 |
| rain | [reɪn] | 瑞恩 | 下雨；雨 |
| season | [ˈsiːzn] | 西镇 | 季节，时期 |
| mouldy | [ˈməʊldi] | 莫尔迪 | 发霉的，旧式的 |
| really | [ˈrɪəli] | 瑞尔丽 | 真的（用来表示兴趣、惊奇或疑惑） |
| mould | [məʊld] | 莫尔德 | 霉，霉菌 |
| bad | [bæd] | 拜德 | 严重的，坏的 |
| air conditioner | [eə] [kənˈdɪʃənə] | 爱尔 肯定迅那 | 空调 |
| dehumidifier | [ˌdiːhjuːˈmɪdɪfaɪə] | 迪休弥迪发尔 | 干燥器，除湿器 |
| on | [ɒn] | 昂 | （电器或电力供应）开着的 |
| everything | [ˈevrɪθɪŋ] | 爱维瑞新 | 每件事物，一切 |
| delay | [dɪˈleɪ] | 迪类 | 耽搁，延期 |
| but | [bʌt] | 拔特 | 但是，然而 |
| suggestion | [səˈdʒestʃ(ə)n] | 色杰思群 | 建议，示意 |
| take | [teɪk] | 忒课 | 接受，拿，取 |
| advice | [ədˈvaɪs] | 阿德外思 | 建议，忠告 |
| overcoat | [ˈəʊvəkəʊt] | 欧无扣特 | 大衣，外套 |
| suit | [sjuːt] | 思由特 | 西装，套装 |

## 家政实用英语

| 英语单词 | 音标 | 发音参考 | 中文翻译 |
|---|---|---|---|
| usually | [ˈjuːʒəli] | 由迅尔丽 | 通常，经常 |
| send | [send] | 三得 | 发动，寄 |
| out | [aʊt] | 奥特 | 外面的，出局的 |
| dry | [draɪ] | 拽尔 | 干的，口渴的 |
| jacket | [ˈdʒækɪt] | 杰克特 | 夹克，短上衣 |
| shirt | [ʃɜːt] | 肖特 | 衬衫，汗衫 |
| pants | [pænts] | 潘次 | 裤子 |
| jeans | [dʒiːnz] | 金子 | 牛仔裤 |
| they | [ðeɪ] | 贼 | 他们，她们，它们 |
| bedding | [ˈbedɪŋ] | 贝丁 | 床上用品 |
| hotel | [həʊˈtel] | 侯太尔 | 酒店，旅馆 |
| husband | [ˈhʌzbənd] | 哈子本德 | 丈夫 |
| better | [ˈbetə] | 倍特 | 更好的，较好的 |
| especially | [ɪˈspeʃ(ə)li] | 一思拜迅丽 | 尤其，特别 |
| collar | [ˈkɒlə] | 考勒 | 衣领 |
| cuff | [kʌf] | 卡夫 | 袖口 |
| him | [hɪm] | 黑幕 | 他 |
| look | [lʊk] | 路克 | 看起来，注意 |
| smart | [smɑːt] | 司马特 | 整洁的，衣着得体的，聪明的 |
| more | [mɔː] | 莫 | 更多；更多的 |
| just | [dʒʌst] | 加思特 | 只是，仅仅 |
| star | [stɑː] | 思达 | 星，恒星，明星 |
| keep | [kiːp] | 克一普 | 保持，经营 |
| standard | [ˈstændəd] | 思单得的 | 标准；标准的 |

 **Useful Sentences** 实用例句

I am planning to do this after the rain season. 我打算在雨季以后再整理收纳。

I will pay more attention. 我以后多注意一点儿。

衣物及洗烫常用语

## Grammar 语法

**一般将来时**：表示将来要发生的事情，在人称代词后面用 will（缩略为 'll）表示，也可以用 be going to 表示将要。

例句：
I will come tomorrow. 我明天会来。
Will you come next week? 你下周来吗?
I am not going to cook dinner today. 我今天不做晚饭。
Summer will come again. 夏天又要来了。
It is going to rain. 要下雨了。
Are you going to school tomorrow? 你明天去学校吗?

## Extension 扩展

常见衣服及配饰

| 英语单词 | 音标 | 发音参考 | 中文翻译 |
| --- | --- | --- | --- |
| hat | [hæt] | 海特 | 帽子 |
| scarf | [skɑːf] | 思卡夫 | 围巾 |
| buttons | [bʌtəns] | 巴腾思 | 纽扣（复数） |
| blouse | [ˈblaʊz] | 博劳兹 | 女装衬衫，宽松上衣 |
| pocket | [ˈpɒkɪt] | 坡克特 | 口袋 |
| belt | [belt] | 贝尔特 | 腰带 |
| shoes | [ˈʃuːz] | 修字 | 鞋子（复数） |
| jumper | [ˈdʒʌmpə] | 江珀 | 套头外衣 |
| tie | [taɪ] | 太 | 领带 |

## Exercises 练习

Dialogue Practice　对话练习

1. Hi Lisa, could you please iron the shirts better?

2. Does it really rain for 1 month during rain season in Shanghai?
3. May I have some suggestions for you?
4. Could you please tell me what you will cook tonight?
5. You clean very well.

**Possible Answers 参考答案**

1. Hi Lisa, could you please iron the shirts better?   — OK.
2. Does it really rain for 1 month during rain season in Shanghai?   — Yes.
3. May I have some suggestions for you?   — Of course, I am happy to listen.
4. Could you please tell me what you will cook tonight?   — ×××.
5. You clean very well.   — Thank you.

# Unit 7
# 第 7 课

烹饪及餐饮礼仪用语
Cooking & Dinning

**Dialogue** 情景对话

Emily: Hi, Lisa, I would like to invite friends for dinner tomorrow. Can you please help with cooking?

艾米莉：嗨，莉萨，我明晚要邀请朋友来吃晚饭。你能帮忙做晚饭吗？

Lisa: Yes, what cuisine would you like to cook?

莉萨：当然，你要做什么菜系？

Emily: My home cuisine, American food would be good. If you can help with some Chinese dishes, that would be very helpful.

艾米莉：我的老家菜，美国菜就好了。如果你可以帮忙做一些中国菜，就太棒了。

Lisa: OK. Then we need to make a shopping list first.

莉萨：好的。那我们先列一张购物清单。

Emily: Oh, yes, I have already made a shopping list: vegetables like tomatoes, potatoes, peas, carrots, broccolis, cucumbers, and mushrooms; fruits including lemons, apples, oranges, grapes, pineapples, and cherries; meats including beef, lamb, pork, and chicken; seafood such as salmon and shrimp.

艾米莉：哦，好的。我已经列了一张购物清单：蔬菜如番茄、土豆、豆子、胡萝卜、西兰花、黄瓜、蘑菇，水果要柠檬、苹果、橙子、葡萄、菠萝、樱桃，肉要牛肉、羊肉、猪肉、鸡肉，海鲜类诸如三文鱼和虾。

Lisa: OK, a lot to buy. I would like to cook Gongbao chicken (spicy diced chicken with peanuts), fish-flavored pork slice and baby lobster. Is it OK?

莉萨：好的，好多东西都要买。我要做宫保鸡丁（辣炒鸡丁和花生）、鱼香肉丝和小龙虾。可以吗？

Emily: Of course, I love these dishes. But please do not cook too hot, add less oil, less salt and less sugar. My friends like healthy food.

艾米莉：当然了，我喜欢这些菜。但是不要做得太辣了，少加点儿油，少点儿盐，少点儿糖。我朋友都喜欢健康食品。

Lisa: OK. I know the Western flavour. I will set the table in the Western style and talk quietly tomorrow.

莉萨：好的。我知道你们西方口味。我会按照西式标准摆盘，而且明天我说话会小声一点儿。

Emily: Thank you. Let's go shopping now.

艾米莉：谢谢。现在我们去买东西吧。

Lisa: Let's go.

莉萨：我们走吧。

 **New Words 新单词**

| 英语单词 | 音标 | 发音参考 | 中文翻译 |
|---|---|---|---|
| cooking | [ˈkʊkɪŋ] | 库金 | 烹饪 |
| dining | [daɪnɪŋ] | 带宁 | 吃饭，进餐 |
| invite | [ɪnˈvaɪt] | 因外特 | 邀请，招待 |
| friend | [frend] | 福然德 | 朋友 |
| dinner | [ˈdɪnə] | 迪拿 | 晚餐，晚宴 |
| cuisine | [kwɪˈziːn] | 快字因 | 菜系，烹饪 |
| American | [əˈmerɪkən] | 阿美瑞肯 | 美国的；美国人 |
| dish | [dɪʃ] | 迪虚 | 烹制好的菜肴，食品，一道菜，盘子，餐具 |
| helpful | [ˈhelpfʊl] | 黑尔普佛 | 有帮助的，有益的 |
| make | [meɪk] | 梅课 | 整理，制造，布置 |
| list | [lɪst] | 立思特 | 列表，清单 |
| vegetable | [ˈvedʒtəbəl] | 饭急特饱 | 蔬菜，植物 |
| already | [ɔːlˈredi] | 奥瑞迪 | 已经，早已 |
| tomato | [təˈmɑːtəʊ] | 拖马透 | 番茄，西红柿 |
| potato | [pəˈteɪtəʊ] | 普忒透 | 土豆，马铃薯 |
| pea | [piː] | 胚芽 | 豌豆 |
| carrot | [ˈkærət] | 开瑞特 | 胡萝卜 |
| broccoli | [ˈbrɒkəli] | 布罗克丽 | 西兰花 |
| cucumber | [ˈkjuːkʌmbə] | 秋肯布 | 黄瓜 |
| mushroom | [ˈmʌʃruːm] | 马旭容 | 蘑菇 |
| fruit | [fruːt] | 福入特 | 水果，果实 |

| 英语单词 | 音标 | 发音参考 | 中文翻译 |
|---|---|---|---|
| lemon | [ˈlemən] | 莱蒙 | 柠檬 |
| apple | [ˈæp(ə)l] | 爱婆儿 | 苹果 |
| orange | [ˈɒrɪn(d)ʒ] | 奥瑞桔 | 橙子，橘子；橙色的 |
| grapes | [greɪps] | 格瑞普思 | 葡萄（复数） |
| pineapple | [ˈpaɪnæpl] | 盼爱婆 | 菠萝，凤梨 |
| meat | [miːt] | 米特 | 肉，肉类 |
| beef | [biːf] | 比夫 | 牛肉 |
| lamb | [læm] | 兰母 | 羔羊肉，羔羊 |
| pork | [pɔːk] | 波尔克 | 猪肉 |
| chicken | [ˈtʃɪkɪn] | 切肯 | 鸡肉，小鸡 |
| seafood | [ˈsiːfuːd] | 西富特 | 海鲜，海产 |
| salmon | [ˈsæmən] | 三门 | 三文鱼，大马哈鱼肉，鲑鱼 |
| shrimp | [ʃrɪmp] | 迅普 | 虾 |
| lot | [lɒt] | 劳特 | 许多，大量 |
| buy | [baɪ] | 拜 | 买，采购 |
| spicy | [ˈspaɪsi] | 思派西 | 辣的，辛香的 |
| diced | [daɪst] | 戴思特 | 切粒的，切成小方块的 |
| peanut | [ˈpiːnʌt] | 皮那特 | 花生 |
| fish | [fɪʃ] | 飞虚 | 鱼，鱼类；捕鱼 |
| slice | [slaɪs] | 思来思 | 薄片；切薄片 |
| baby | [ˈbeɪbi] | 北鼻 | 婴儿，孩子气的人 |
| lobster | [ˈlɒbstə] | 罗布斯特 | 龙虾，龙虾肉 |
| hot | [hɒt] | 好特 | 热的，辣的 |
| add | [æd] | 爱德 | 增加，添加 |
| less | [les] | 莱斯 | 较少的，较小的 |
| oil | [ɒɪl] | 欧尔欧 | 油，石油 |
| salt | [sɔːlt] | 扫尔特 | 盐；咸的 |
| sugar | [ˈʃʊgə] | 需格尔 | 糖，食用糖 |
| healthy | [ˈhelθi] | 黑尔西 | 健康的，健全的 |
| flavor | [ˈfleɪvə] | 富蕾五 | 口味，香味 |

烹饪及餐饮礼仪用语

| 英语单词 | 音标 | 发音参考 | 中文翻译 |
|---|---|---|---|
| set | [set] | 赛特 | 放，置；集合，一套 |
| table | [ˈteɪb(ə)l] | 忒布 | 桌子，表格 |
| style | [staɪl] | 思戴尔 | 风格，类型 |

## Useful Sentences　实用例句

What cuisine would you like to cook?　你要烧什么菜系？

If you can help with some Chinese dishes, that would be very helpful.
如果你可以帮忙做一些中国菜，就太棒了。

Let's go.　我们走吧。

## Grammar　语法

一般过去时：表示一个过去发生的动作或存在的状态，也可以表示过去习惯性的动作，刚发生的动作也要用过去时表示，在叙述性文字中，如小说、故事等也常用过去时。

例句：

I made a shopping list. 我列了一张购物清单。

They loved each other for the past 10 years. 在过去的10年里他们彼此相爱。

She was a cook. 她曾是个厨师。

He was here 5 minutes ago. 他5分钟前在这儿。

Newton was a great scientist. 牛顿是位伟大的科学家。

## Extension　扩展

水果蔬菜

| 英语单词 | 音标 | 发音参考 | 中文翻译 |
|---|---|---|---|
| banana | [bəˈnɑːnə] | 巴娜娜 | 香蕉 |
| cherry | [ˈtʃeri] | 切瑞 | 樱桃 |

**家政实用英语**

| 英语单词 | 音标 | 发音参考 | 中文翻译 |
|---|---|---|---|
| watermelon | [ˈwɔːtəmelən] | 我特没冷 | 西瓜 |
| cabbage | [ˈkæbɪdʒ] | 凯比奇 | 卷心菜 |
| onion | [ˈʌnjən] | 阿念 | 洋葱 |
| cauliflower | [ˈkɒlɪflauə] | 扣丽福老儿 | 花椰菜，菜花 |

**Exercises 练习**

Dialogue Practice   对话练习

1. Do you mind if I invite friends over for dinner tomorrow?
2. What dishes would you like me to cook tomorrow?
3. What do you need me to buy for tomorrow?
4. How many dishes shall I prepare?
5. Shall I cook Chinese or Western food?

**Possible Answers 参考答案**

1. Do you mind if I invite friends over for dinner tomorrow?    — Of course not, they are welcome.

2. What dishes would you like me to cook tomorrow?    — ×××.

3. What do you need me to buy for tomorrow?    — ×××.

4. How many dishes shall I prepare?    — 5-10 dishes are good.

5. Shall I cook Chinese or Western food?    — Some Chinese and some Western dishes.

# Unit 8
# 第 8 课

房屋设备及家用电器用语
Housing Facilities & Home Appliances

家政实用英语

**Dialogue** 情景对话

Lisa： Good morning, Emily. Could you please show me how to use home appliances?

莉萨： 早上好，艾米莉。你能教我一下怎么用家用电器吗？

Emily： OK. Let's start from kitchen. This is oven for baking, fridge, dish washer and garbage crusher.

艾米莉： 好的，我们先从厨房开始。这是烘焙用的烤箱、冰箱、洗碗机和垃圾粉碎机。

Lisa： I like dish washer, it saves lots of time.

莉萨： 我喜欢洗碗机，省了很多时间。

Emily： I agree. Do you know how to use the garbage crusher?

艾米莉： 我也这么觉得。你知道怎么用垃圾粉碎机吗？

Lisa： Oh, yes, my previous employer also had one, same brand.

莉萨： 是的，我知道的，我上个雇主也有一个，同一个品牌的。

Emily： Great. This is coffee machine, and water dispenser.

艾米莉： 太棒了。这是咖啡机和净水器。

Lisa： Yes, cook with purified water, right?

莉萨： 好的，做饭用净水，对吗？

Emily： Yes, correct. And here comes the vacuum cleaner and iron.

艾米莉： 是的。这是吸尘器和熨斗。

Lisa： OK, I know how to use, I iron very well.

莉萨： 好的，我知道怎么用，我熨烫得很好。

Emily： Good to know this. Here are the remote controllers for air conditioners and TV.

艾米莉： 很高兴知道你熨烫得好。这是空调和电视机的遥控器。

Lisa： Shall I keep air conditioner on when you are not at home?

莉萨： 你们不在家的时候，我也要开着空调吗？

Emily： No need if nobody is at home. If you have any question about any home appliances or housing facilities, please let me know anytime.

艾米莉： 如果没人在家不用开着。如果对于家用电器或者房屋设备你还有问题，随时来问我。

Lisa： OK, thank you!

莉萨： 好的，谢谢！

## 房屋设备及家用电器用语

### New Words 新单词

| 英语单词 | 音标 | 发音参考 | 中文翻译 |
| --- | --- | --- | --- |
| housing | [ˈhaʊzɪn] | 好增 | 房屋 |
| facilities | [fəˈsɪlɪtɪz] | 福思立体姿 | 设施，工具（复数） |
| appliance | [əˈplaɪəns] | 阿普来恩思 | 器械，器具 |
| me | [miː] | 蜜 | 我，自我 |
| oven | [ˈʌvən] | 阿文 | 烤箱，炉，灶头 |
| baking | [ˈbeɪkɪn] | 贝肯宁 | 烘焙，烘烤 |
| fridge | [frɪdʒ] | 福瑞桔 | 电冰箱 |
| washer | [ˈwɒʃə] | 我洗 | 洗碗机，洗衣服的人 |
| garbage | [ˈɡɑːbɪdʒ] | 加比其 | 垃圾，废物 |
| crusher | [ˈkrʌʃə] | 克鲁虚 | 粉碎机，压碎的东西 |
| save | [seɪv] | 思爱五 | 节省，保存 |
| agree | [əˈɡriː] | 阿格瑞 | 同意，赞成，承认 |
| previous | [ˈpriːvɪəs] | 普瑞维尔思 | 以前的，早先的 |
| employer | [ɪmˈplɔɪə] | 英普罗叶尔 | 雇主，老板 |
| one | [wʌn] | 万安 | 一个人，任何人 |
| brand | [brænd] | 布然的 | 商标，牌子 |
| coffee | [ˈkɒfi] | 咖啡 | 咖啡，咖啡豆，咖啡色 |
| machine | [məˈʃiːn] | 么迅 | 机器，机械 |
| water | [ˈwɔːtə] | 沃特 | 水 |
| dispenser | [dɪˈspensə] | 迪思潘色 | 分配者，自动售货机，饮水机 |
| purified | [ˈpjʊərɪfaɪd] | 普尔瑞范的 | 净化的，精制的 |
| right | [raɪt] | 瑞特 | 正确的，直接的 |
| correct | [kəˈrekt] | 克瑞克特 | 改正，纠正；正确的 |
| remote | [rɪˈməʊt] | 瑞谋特 | 远程的，遥远的 |
| controller | [kənˈtrəʊlə] | 肯秋乐 | 控制器，管理者 |
| TV | [ˌtiːˈviː] | 提味 | 电视机（television 的缩写） |

**家政实用英语**

| 英语单词 | 音标 | 发音参考 | 中文翻译 |
|---|---|---|---|
| nobody | [ˈnəubədi] | 诺贝迪 | 无人，没有人，没有任何人 |
| any | [ˈeni] | 爱宁 | 任何的，任何 |
| question | [ˈkwestʃən] | 快思群 | 问题，疑问 |
| anytime | [ˈenɪtaɪm] | 爱宁太母 | 任何时候 |

## Useful Sentences 实用例句

Could you please show me how to use the home appliances?
你能教我一下怎么用电器吗？
Here comes the vacuum cleaner and iron. 这是吸尘器和熨斗。

## Grammar 语法

**一般现在时**：表示经常性、现在性、规律性、计划性或者真理性的状态或动作。

例句：

He gets up at 7am every day. 他每天早上7点起床。

She is a housekeeper. 她是位家政服务员。

It snows in winter. 冬天会下雪。

I love ironing. 我爱熨烫。

Here comes the vacuum cleaner and iron. 这是吸尘器和熨斗。

If you have any question about any home appliances or housing facilities, let me know anytime. 如果你对家用电器或家里的设备有疑问，随时告诉我。

## Extension 扩展

厨房用品

## 房屋设备及家用电器用语

| 英语单词 | 音标 | 发音参考 | 中文翻译 |
|---|---|---|---|
| kitchen | [ˈkɪtʃɪn] | 开群 | 厨房，炊具 |
| tap | [tæp] | 塔普 | 水龙头 |
| sink | [sɪŋk] | 新课 | 水槽，洗手池 |
| dishwasher | [ˈdɪʃwɒʃə] | 碟需我洗 | 洗碗机 |
| cupboard | [ˈkʌbəd] | 卡卜得 | 橱柜 |
| kettle | [ˈketəl] | 凯特尔 | 烧水壶 |
| cooker | [ˈkʊkə] | 库克 | 炊具 |

**Exercises 练习**

Dialogue Practice 对话练习

1. Do you know how to use home appliances?
2. I always bake with oven, do you like baking?
3. Could you please make coffee every morning?
4. May I have some suggestion for your cooking?
5. Would you mind if I turn on air conditioner when I clean here?

**Possible Answers 参考答案**

1. Do you know how to use home appliances? — Yes, I know.
2. I always bake with oven, do you like baking? — Yes, I like.
3. Could you please make coffee every morning? — Yes, I will.
4. May I have some suggestion for your cooking? — Yes, of course.
5. Would you mind if I turn on air conditioner when I clean here? — No, not at all.

Unit 9
第 9 课

孩子照护及日常教育常用语
Child Care & Daily Education

### Dialogue 情景对话

Emily: Hi Lisa, Thomas is sleeping now. Please be quiet!
艾米莉: 嗨, 莉萨。托马斯正在睡觉, 请安静一点。

Lisa: OK.
莉萨: 好的。

Emily: I would like to tell you about our rules for Thomas.
艾米莉: 我跟你说一下我们给托马斯的安排。

Lisa: Sure.
莉萨: 好的。

Emily: Here is the schedule for a typical day of Thomas.
艾米莉: 这是托马斯的每日作息安排。

Lisa: OK. Milk at 8am, and then have play and snack time till 9:30am. Afterwards, take him to the play ground to meet other babies, milk again at 11:30am, then take a nap at 12pm.
莉萨: 好的, 早上8点喝奶, 然后玩一会, 吃点心, 9:30以后带托马斯去儿童游乐场和其他小宝宝一起玩, 11:30再喝奶, 然后12点午睡。

Emily: Yes, correct. I need you to follow the schedule well.
艾米莉: 是的, 对的。我需要你严格按照时间表来做。

Lisa: OK. If Thomas doesn't want to sleep, what shall I do?
莉萨: 好的, 如果托马斯不想睡觉, 我怎么办呢?

Emily: Please put him to bed, tell him to sleep and close the door. If he doesn't cry, please leave him alone.
艾米莉: 把托马斯放在床上, 告诉他要睡觉了, 然后把门关上。如果他不哭, 那就不用管了。

Lisa: Sure.
莉萨: 好的。

Emily: I need you to follow my rules strictly, no exception, only when he is sick.
艾米莉: 我需要你严格遵守规则, 不能例外, 只有当孩子生病的时候才能有例外。

Lisa: Yes, I will.
莉萨: 好的, 我知道的。

Emily: Sing him a song and read him a story before he sleeps.

艾米莉:睡觉之前,给他唱首歌,读个故事。

Lisa: OK, I know many songs.

莉萨:好的,我会唱很多歌。

Emily: You need to discipline Thomas but no shouting.

艾米莉:你需要让托马斯守规矩,但是不能冲着孩子吼。

Lisa: Sure, I love children and I would always talk gently.

莉萨:好的,我喜欢孩子,一直跟孩子轻轻说话。

Emily: Good, thank you.

艾米莉:很好,谢谢你。

Lisa: My pleasure.

莉萨:这是我的荣幸。

## New Words 新单词

| 英语单词 | 音标 | 发音参考 | 中文翻译 |
| --- | --- | --- | --- |
| education | [ˌedjʊˈkeɪʃən] | 爱丢开勋 | 教育,培养 |
| sleep | [sliːp] | 思丽普 | 睡觉 |
| quiet | [ˈkwaɪət] | 块尔特 | 安静的,安定的 |
| rule | [ruːl] | 入耳 | 规则;统治,管理 |
| typical | [ˈtɪpɪkl] | 提皮克尔 | 典型的,特有的 |
| milk | [mɪlk] | 谬尔克 | 牛奶,乳状物 |
| snack | [snæk] | 思耐克 | 小吃,快餐 |
| afterwards | [ˈɑːftəwədz] | 阿福特我自 | 然后,后来 |
| other | [ˈʌðə] | 阿泽 | 其他的,另外的 |
| again | [əˈgen] | 阿甘 | 再一次,此外,又 |
| nap | [næp] | 耐普 | 小睡,打盹儿 |
| follow | [ˈfɒləʊ] | 佛楼 | 跟随,遵循 |
| close | [kləʊs] | 克楼姿 | 紧密的,亲近的 |
| door | [dɔː] | 夺 | 门,户 |
| cry | [kraɪ] | 克来 | 哭,叫喊 |
| alone | [əˈləʊn] | 阿尔隆 | 独自的,单独的 |

家政实用英语

| 英语单词 | 音标 | 发音参考 | 中文翻译 |
|---|---|---|---|
| strictly | [ˈstrɪktli] | 思坠克特雷 | 严格地，完全地 |
| exception | [ɪkˈsepʃ(ə)n] | 一克塞普迅 | 例外，异议 |
| only | [ˈəʊnli] | 欧雷 | 仅仅，只有 |
| sick | [sɪk] | 思尔克 | 生病的，不舒服的；病人 |
| sing | [sɪŋ] | 新 | 唱歌，歌颂，鸣叫 |
| song | [sɒŋ] | 桑 | 歌曲，歌唱 |
| read | [riːd] | 瑞德 | 阅读 |
| story | [ˈstɔːri] | 思道瑞 | 故事，小说 |
| many | [ˈmeni] | 曼宁 | 许多，许多人 |
| shouting | [ˈʃaʊtɪŋ] | 消停 | 大喊大叫 |
| talk | [tɔːk] | 套克 | 说，谈话，讨论 |
| gently | [ˈdʒentli] | 占特立 | 轻轻地，温柔地 |
| pleasure | [ˈpleʒə] | 普来学 | 快乐，娱乐，使高兴 |

### Useful Sentences 实用例句

I need you to follow the schedule well.
我需要你严格按照时间表来做。

If Thomas doesn't want to sleep, what shall I do?
如果托马斯不想睡觉，我怎么办呢？

### Grammar 语法

**现在进行时**：表示现在正在进行的动作、发生的事情或存在的状态。一般在动词后面加 ing 表示现在正在进行。现在进行时有时候也可以表示将来的动作，一般跟着一个表示未来的时间词，表示安排好的事情。

> 例句：
> Thomas is sleeping now. 托马斯在睡觉。
> What are you doing? 你在做什么？
> It's raining outside. 外面在下雨。

I am writing a book about English for housekeepers. 我在写一本关于家政英语的书。

I am going shopping this afternoon. 我下午去采购。

 Extension 扩展

| Age 年龄（years 岁） | Stage 阶段 |
| --- | --- |
| 0~1 | baby 婴儿 |
| 1~2 | toddler 幼童，学步的孩子 |
| 2~12 | child 儿童 |
| 13~17 | teenager 青少年 |
| 18+ | adult 成年人 |

 Exercises 练习

Dialogue Practice 对话练习

1. This is the schedule for Thomas everyday, I need you to follow the schedule well. Is it OK for you?

2. What kind of baby song can you sing?

3. Do you know how to make baby food?

4. What baby food can you make?

5. Do you know how to talk to baby?

 Possible Answers 参考答案

1. This is the schedule for Thomas everyday, I need you to follow the schedule well. Is it OK for you?　　—— OK, I will follow the schedule well.

2. What kind of baby song can you sing?　　—— Jingle bell, 3 apples, etc.

3. Do you know how to make baby food?　　—— Yes, I know very well.

4. What baby food can you make?　　—— Rice cereal, smashed potato, steamed egg, etc.

5. Do you know how to talk to baby?　　—— Yes, I know, talk gently.

ns and Plants
# Unit 10
## 第 10 课

宠物及相关动植物英文名称
Pets, Animals & Plants

## Dialogue 情景对话

Emily: Hi, Lisa, do you like pets?
艾米莉：嗨，莉萨，你喜欢宠物吗？

Lisa: Yes, I like dogs most, cats, rabbits, hamsters are also OK.
莉萨：是的，我最喜欢狗，猫、兔子、仓鼠也挺喜欢。

Emily: Good to know this. We will have a dog and two cats coming in two weeks' time.
艾米莉：太好了，我们会有一条狗和2只猫，2周以后接回来。

Lisa: Oh, good, I know how to take care of dogs and cats. I like walking dogs, it's very good exercise. I have rich experience taking care of pets.
莉萨：好的，非常好。我知道怎么照顾猫和狗。我喜欢遛狗，这是很好的锻炼。我有照顾宠物的丰富经验。

Emily: Great. I have also bought some plants, including red roses, basils, mints, a little pine tree for Christmas, and cactus. They will be delivered tomorrow.
艾米莉：太棒了。我也买了一些植物，包括一些红玫瑰、罗勒、薄荷、一棵小松树、仙人掌，明天送到家里来。

Lisa: OK. Could you please tell me how often I shall water the plants?
莉萨：好的，你能告诉我多久浇一次水吗？

Emily: Sure, there is a note on each plant. Some need to be watered once a week, some twice a week, and some just once a month, you just need to follow the note.
艾米莉：好的，每盆花上面都有一个标签，有些一周浇一次水，有些一周浇两次水，有些一个月浇一次，你只需要根据标签来浇水就可以了。

Lisa: OK. Anything else I shall pay attention to?
莉萨：好的，还有什么其他我需要注意的吗？

Emily: That's all for now. Please just keep my plants alive. Thank you!
艾米莉：暂时就这些了。只要保持我的植物活着。谢谢！

Lisa: My pleasure.
莉萨：这是我的荣幸。

宠物及相关动植物英文名称

 New Words  新单词

| 英语单词 | 音标 | 发音参考 | 中文翻译 |
| --- | --- | --- | --- |
| animal | [ˈænɪməl] | 爱你莫 | 动物；动物的 |
| pet | [pet] | 派特 | 宠物 |
| dog | [dɒg] | 道格 | 狗 |
| most | [məʊst] | 谋思特 | 大部分的，多数的 |
| cat | [kæt] | 凯特 | 猫，猫科动物 |
| rabbit | [ˈræbɪt] | 瑞比特 | 兔子，野兔 |
| hamster | [ˈhæmstə] | 汉姆斯特 | 仓鼠 |
| two | [tuː] | 吐 | 二，两个 |
| walk | [wɔːk] | 卧克 | 步行，散步 |
| exercise | [ˈeksəsaɪz] | 爱克色赛姿 | 运动，练习 |
| rich | [rɪtʃ] | 瑞驰 | 丰富的，富有的，肥沃的 |
| experience | [ɪkˈspɪərɪəns] | 一克思皮尔润思 | 经验，经历 |
| plant | [plɑːnt] | 普朗特 | 植物；种植 |
| red | [red] | 瑞德 | 红色；红色的 |
| rose | [rəʊz] | 肉姿 | 玫瑰，粉红色 |
| basil | [ˈbæzəl] | 白姿尔 | 罗勒 |
| mint | [mɪnt] | 民特 | 薄荷 |
| little | [ˈlɪtəl] | 力特尔 | 小的，很少的，短暂的 |
| pine | [paɪn] | 潘安 | 松树，凤梨 |
| tree | [triː] | 特瑞 | 树，木料 |
| Christmas | [ˈkrɪsməs] | 克瑞思莫斯 | 圣诞节 |
| cactus | [ˈkæktəs] | 凯克特思 | 仙人掌 |
| deliver | [dɪˈlɪvə] | 迪立无 | 递送，交付 |
| often | [ˈɒfn] | 欧芬 | 经常，时常 |
| note | [nəʊt] | 诺特 | 笔记，注解 |
| each | [iːtʃ] | 亿趣 | 每个，各自 |

家政实用英语

| 英语单词 | 音标 | 发音参考 | 中文翻译 |
|---|---|---|---|
| once | [wʌns] | 玩思 | 一次，曾经 |
| twice | [twaɪs] | 特外思 | 两次，两倍 |
| alive | [əˈlaɪv] | 阿来五 | 活着的，活泼的 |

## Useful Sentences 实用例句

Do you like pets?　你喜欢宠物吗？

I like dogs most, cats, rabbits, hamsters are also OK.

我最喜欢狗，猫、兔子、仓鼠也挺喜欢。

## Grammar 语法

**形容词**：用来描写或修饰名词或代词，表示人或事物的品质、状态、特征或属性的词汇，例如：nice, beautiful, happy, rich 等。有些形容词除了原级外，还有两种形式，即比较级和最高级。比较级表示"比较……"，最高级表示"最……"，一般单音节单词的比较级在单词后面加 er，最高级在单词后面加 est；双音节或者多音节单词在前面加 more 和 most 来构成比较级和最高级。

| 原级 | 比较级 | 最高级 |  |
|---|---|---|---|
| rich | richer | richest |  |
| nice | nicer | nicest | （以 e 结尾，直接加 r 和 st） |
| beautiful | more beautiful | most beautiful |  |
| happy | happier | happiest | （以 y 结尾，变 y 为 i，再加 er 和 est） |

还有些形容词有不规则的比较级和最高级形式。

| 原级 | 比较级 | 最高级 |
|---|---|---|
| good, well | better | best |
| bad | worse | worst |
| much, many | more | most |
| little | less | least |
| far | farther, further | farthest, furthest |

宠物及相关动植物英文名称

Extension  扩展

常见动物和昆虫

| 英语单词 | 音标 | 发音参考 | 中文翻译 |
|---|---|---|---|
| elephant | [ˈelɪfənt] | 爱丽分特 | 大象 |
| bear | [beə] | 贝尔 | 熊 |
| lion | [ˈlaɪən] | 来恩 | 狮子 |
| tiger | [ˈtaɪgə] | 泰格 | 老虎 |
| monkey | [ˈmʌŋki] | 芒科 | 猴子 |
| zebra | [ˈziːbrə] | 自不拉 | 斑马 |
| bee | [biː] | 毕 | 蜜蜂 |
| butterfly | [ˈbʌtəflaɪ] | 巴特福莱 | 蝴蝶 |
| fish | [fɪʃ] | 飞虚 | 鱼，鱼类，捕鱼 |
| shark | [ʃɑːk] | 沙克 | 鲨鱼 |

Exercises  练习

Dialogue Practice  对话练习

1. Do you like pets?

2. What pets do you like?

3. Shall I walk the dog 3 times a day?

4. How often do you need me to water the plants?

5. Can you come once a week to water the plants during summer holiday?

Possible Answers  参考答案

1. Do you like pets?　　— Yes, I like pets.

2. What pets do you like?　　— I like dogs, cats, birds.

3. Shall I walk the dog 3 times a day?　　— Yes, thank you.

4. How often do you need me to water the plants?　　— Once a week is good.

5. Can you come once a week to water the plants during summer holiday?　　— Yes, I can.

# Unit 11
# 第 11 课

家庭聚会及社交活动用语
Family Party & Social Activities

### Dialogue 情景对话

Lily: Hey, mum, are we going back to hometown for Christmas?

莉莉：嘿，妈妈，我们圣诞节回老家吗？

Emily: Of course. Grandparents miss us a lot. And we will have a family party then, to meet all our relatives and friends.

艾米莉：当然。祖父母很想念我们。我们回家之后要开一个家庭聚会，和所有亲戚、朋友聚聚。

Lily: Great. When will we leave and come back?

莉莉：太棒了。我们什么时候走，什么时候回来？

Emily: About 15th December, then come back around 4th January.

艾米莉：大约12月15日走，1月4日前后回来。

Lily: Regarding the party, may I also invite my friends?

莉莉：关于那个聚会，我也可以邀请我的朋友们吗？

Emily: Sure, you may invite whomever you like.

艾米莉：当然，你可以邀请任何你喜欢的人。

Lily: Can we also go skiing during Christmas?

莉莉：我们圣诞节也能去滑雪吗？

Emily: Yes, we are still planning for this trip. Maybe we can invite Richard's family together for skiing. Your father had very good relationship with Richard before, he was promoted when we came to China.

艾米莉：是的，我们还在规划旅行。或许我们可以邀请理查德一家去滑雪。你爸爸和他们家关系很好，我们来中国以后，理查德就升职了。

Lily: Oh, mum, you can't be so snobbish.

莉莉：哦，妈妈，你可不能这么势利。

Emily: Well, don't you want to meet Richard's daughter Tina? You were good friends before.

艾米莉：嗯，你不想见见理查德家的女儿蒂娜吗？你们以前可是好朋友。

Lily: Oh, yes. I quite like her. Can we also invite our ayi Lisa to go with us?

莉莉：哦，当然，我很喜欢她。我们也能邀请阿姨莉萨一起去吗？

Emily: Sure, if she likes, we would definitely invite her to come with us. Or she can have days off when we are gone. She may want to be with her family.

艾米莉：当然，如果她想去，我们当然可以邀请她一起去。或者我们不在的时候她也可以放假，她或许想跟家人在一起。

Lily：Oh, great, I will tell Lisa.

莉莉：太棒了，我去告诉莉萨。

New Words  新单词

| 英语单词 | 音标 | 发音参考 | 中文翻译 |
| --- | --- | --- | --- |
| social | [ˈsəuʃəl] | 搜旭 | 社交的，社会的 |
| activity | [ækˈtɪvəti] | 艾克提吴提 | 活动，行动 |
| mum | [mʌm] | 妈姆 | 妈妈 |
| back | [bæk] | 白客 | 回原处；后面，背部 |
| hometown | [ˈhəumˈtaun] | 厚木唐 | 家乡，故乡 |
| grandparents | [ˈgrændpeərənts] | 格兰德派尔润次 | 祖父母，外祖父母 |
| miss | [mɪs] | 密斯 | 思念，留恋 |
| family | [ˈfæmɪli] | 饭米粒 | 家庭，家属，家族 |
| party | [ˈpɑːti] | 趴体 | 派对，聚会，政党 |
| relative | [ˈrelətɪv] | 瑞乐提五 | 亲戚；相关的 |
| December | [dɪˈsembə] | 迪三部 | 十二月 |
| January | [ˈdʒænjuəri] | 占扭尔瑞 | 一月 |
| regarding | [rɪˈgɑːdɪŋ] | 瑞卡丁 | 关于 |
| whomever | [huːmˈevə] | 胡母爱维 | 无论谁 |
| skiing | [ˈskiːɪŋ] | 思克英 | 滑雪 |
| during | [ˈdjuərɪŋ] | 丢尔瑞英 | 在……的时候 |
| still | [stɪl] | 思丢尔 | 仍然，尽管如此 |
| trip | [trɪp] | 特瑞普 | 旅行 |
| maybe | [ˈmeɪbiː] | 梅币 | 也许，可能 |
| together | [təˈgeðə] | 特盖着 | 一起，同时，相互 |
| father | [ˈfɑːðə] | 发泽 | 父亲，爸爸 |
| relationship | [rɪˈleɪʃənʃɪp] | 瑞雷迅学普 | 关系，关联 |
| promote | [prəˈməut] | 普瑞谋特 | 促进，提升，推销 |
| China | [tʃaɪnə] | 拆拿 | 中国 |

## 家政实用英语

| 英语单词 | 音标 | 发音参考 | 中文翻译 |
|---|---|---|---|
| so | [səʊ] | 搜 | 如此,这么 |
| snobbish | [ˈsnɒbɪʃ] | 思诺比洗 | 势利的 |
| daughter | [ˈdɔːtə] | 多特 | 女儿 |
| quite | [kwaɪt] | 快特 | 很,相当,完全 |
| she | [ʃiː] | 洗 | 她 |
| definitely | [ˈdefɪnɪtli] | 叠非倪特丽 | 明确地,清楚地 |
| or | [ɔː] | 奥 | 或,或者 |
| off | [ɒf] | 奥芙 | 离开,切断,走开 |
| gone | [gɒn] | 刚 | 离去的,用光的 |
| may | [meɪ] | 梅 | 也许,可能 |
| her | [hɜː] | 贺 | 她,她的 |

### Useful Sentences 实用例句

You can invite whomever you like. 你可以邀请任何你喜欢的人。

Don't you want to meet Richard's daughter Tina? 你不想见见理查德家的女儿蒂娜吗?

### Grammar 语法

**序数词**:表示数目顺序,除了"第一""第二""第三"外,其他序数词都以在基数词后面加词尾 th 构成,有些特殊变形的,已用粗体表示。两位数的序数词,只需要把个位数变为序数词即可。三位以上的序数词,只要把最后两位数变为序数词即可。

| 基数词 | 序数词 | 缩写 | 基数词 | 序数词 | 缩写 | 基数词 | 序数词 | 缩写 |
|---|---|---|---|---|---|---|---|---|
| **1** | **first** | **1st** | 10 | tenth | 10th | 19 | nineteenth | 19th |
| **2** | **second** | **2nd** | 11 | eleventh | 11th | **20** | **twentieth** | **20th** |
| **3** | **third** | **3rd** | **12** | **twelfth** | **12th** | **30** | **thirtieth** | **30th** |
| 4 | fourth | 4th | 13 | thirteenth | 13th | **40** | **fortieth** | **40th** |
| **5** | **fifth** | **5th** | 14 | fourteenth | 14th | **50** | **fiftieth** | **50th** |
| 6 | sixth | 6th | 15 | fifteenth | 15th | **60** | **sixtieth** | **60th** |
| 7 | seventh | 7th | 16 | sixteenth | 16th | **70** | **seventieth** | **70th** |
| **8** | **eighth** | **8th** | 17 | seventeenth | 17th | **80** | **eightieth** | **80th** |
| **9** | **ninth** | **9th** | 18 | eighteenth | 18th | **90** | **ninetieth** | **90th** |

例句：

We live on the ninth floor. 我们住在九楼。

Thomas is our second son. 托马斯是我们二儿子。

We will leave on 15th December. 我们12月15日走。

She has a second hand car. 她有一辆二手车。

Is this your first visit to China? 这是你第一次来中国吗？

Tomorrow is my 16th birthday. 明天是我16岁生日。

## Extension 扩展

一年的12个月

| January | 一月 | February | 二月 |
| March | 三月 | April | 四月 |
| May | 五月 | June | 六月 |
| July | 七月 | August | 八月 |
| September | 九月 | October | 十月 |
| November | 十一月 | December | 十二月 |

## Exercises 练习

1. We will have a party this Saturday, could you please come to help?

2. Could you please help me to write Chinese on the invitation card?

3. What would you like me to prepare for the party?

4. Hi Lisa, if your husband and son are free this Saturday, they are also welcome to the party.

5. Do we have enough wine and soft drinks for the party?

## Possible Answers 参考答案

1. We will have a party this Saturday, could you please come to help?

— Yes, of course.

2. Could you please help me to write Chinese on the invitation card?

— Yes, I am happy to.

3. What would you like me to prepare for the party?

— Some meat, some vegetable and some snack.

4. Hi Lisa, if your husband and son are free this Saturday, they are also welcome to the party.

— Thank you.

5. Do we have enough wine and soft drinks for the party?

— Yes, we have enough.

Unit 12
第 12 课

生理构造及常见疾病名称
Physiology & Common Diseases

### Dialogue 情景对话

Sarah: Hi, everyone, we are going to have a class of health and disease today.

莎拉: 嗨，大家好，今天我们上一节关于健康和疾病的课程。

Children: OK, Sarah!

孩子们: 好的，莎拉。

Sarah: First, let's learn about our body. This is our head, hair, nose, eyes, ears, mouth and neck. When I say the word, please touch the body part. Are you ready?

莎拉: 首先，我们了解一下我们的身体。这是头、头发、鼻子、眼睛、耳朵、嘴巴和脖子。我说单词的时候，请用手点相应的部位。准备好了吗？

Children: Yes, we are ready.

孩子们: 是的，我们准备好了。

Sarah: Please touch your nose, mouth, neck, hair, eyes, ears. Great, all of you do it very well!

莎拉: 请点鼻子、嘴巴、脖子、头发、眼睛、耳朵。太棒了，你们都做得很棒！

Children: Thank you, Sarah!

孩子们: 谢谢，莎拉。

Sarah: Now let's continue with our body, arms, hands, legs and feet. Please touch your body, hands, legs, and feet. All right, you are all doing very well again.

莎拉: 现在我们开始身体部分，手臂、手、腿和脚。点你的身体、手、腿和脚。好了，你们又表现得很棒。

Children: Thanks, Sarah!

孩子们: 谢谢，莎拉。

Sarah: Now let's see what kind of disease we may catch. Cough, sore throat, flu, high fever, and headache.

莎拉: 现在我们来看看我们可能会得哪些疾病。咳嗽、嗓子哑、流感、发烧、头疼。

Jack: I want to be healthy all the time.

杰克: 我希望一直保持健康。

Monica: Me too, but I am coughing today for the bad air.

# 生理构造及常见疾病名称

莫妮卡：我也是，但是因为空气不好，我今天有点咳嗽。

Gordon：I have a little high fever today.

高登：我今天有一点发烧。

Isabel：I have headache today.

伊莎贝尔：我今天有点头疼。

Sarah：Come on, kids, if you are sick, you'd better see our school doctor. You all know where to find the doctor. If not, let's continue with our lesson.

莎拉：孩子们，好了，如果你们谁生病了，最好去看我们的校医，你们都知道医生在哪里。如果没有生病，我们就继续上课。

Children：Hahahahaha.

孩子们：哈哈哈哈哈哈。

Sarah：May I ask one question? Anybody knows how to stay healthy?

莎拉：我能问个问题吗？谁知道怎么保持健康吗？

Children：Wash hands, eat clean food, no smoking, have enough sleep, and be happy.

孩子们：洗手，吃干净的食物，不抽烟，睡眠充足，保持快乐。

Sarah：Great, kids, good to know all these ways, you are so smart to stay healthy. Now class is over, anybody wants to see doctor?

莎拉：太好了，孩子们，很高兴知道你们有这么多方法，你们都很聪明。现在下课，还有谁想要去看医生吗？

Children：No, thank you, Sarah.

孩子们：不了，谢谢，莎拉。

（备注：在英文里称呼老师常用 Sir, Ms, Miss, Mr... 上述对话中 Sarah 可改成 Miss。）

**New Words 新单词**

| 英语单词 | 音标 | 发音参考 | 中文翻译 |
| --- | --- | --- | --- |
| physiology | [ˌfɪzɪˈɒlədʒi] | 菲子奥乐其 | 生理学 |
| common | [ˈkɒmən] | 考门 | 常见的，普通的，一般的 |
| disease | [dɪˈziːz] | 迪西子 | 疾病，病 |
| everyone | [ˈevrɪwʌn] | 艾维瑞万 | 每个人，人人 |
| class | [klɑːs] | 科拉思 | 班级，等级 |
| health | [helθ] | 黑尔思 | 健康，卫生 |

家政实用英语

| 英语单词 | 音标 | 发音参考 | 中文翻译 |
|---|---|---|---|
| illness | [ˈɪlnəs] | 一尔尼斯 | 病，疾病 |
| today | [təˈdeɪ] | 特德 | 今天，现今 |
| teacher | [ˈtiːtʃə] | 提切尔 | 教师，导师 |
| learn | [lɜːn] | 伦恩 | 学习，知道 |
| body | [ˈbɒdi] | 波底 | 身体，主题 |
| head | [hed] | 黑的 | 头，头部 |
| hair | [heə] | 黑尔 | 头发，毛发 |
| nose | [nəʊz] | 诺姿 | 鼻子，嗅觉 |
| eye | [aɪ] | 爱 | 眼睛，视力 |
| ear | [ɪə] | 叶耳 | 耳朵，听觉 |
| mouth | [maʊθ] | 冒思 | 口，嘴 |
| neck | [nek] | 耐克 | 脖子，衣领 |
| say | [seɪ] | 赛尔 | 讲话，说明 |
| word | [wɜːd] | 卧德 | 单词，话语 |
| touch | [tʌtʃ] | 他取 | 接触，触摸 |
| part | [pɑːt] | 怕特 | 部分，角色 |
| ready | [ˈredi] | 瑞德 | 准备好，迅速地 |
| continue | [kənˈtɪnjuː] | 肯提扭 | 继续，延续 |
| arm | [ɑːm] | 阿母 | 手臂，武器 |
| leg | [leg] | 来个 | 腿，支柱 |
| feet | [fiːt] | 菲特 | 脚（复数） |
| kind | [kaɪnd] | 看德 | 和蔼的，宽容的 |
| catch | [kætʃ] | 凯曲 | 赶上，抓住 |
| cough | [kɒf] | 考夫 | 咳嗽，咳嗽声 |
| sore | [sɔː] | 索尔 | 疼痛的，痛心的 |
| throat | [θrəʊt] | 思肉特 | 喉咙，嗓子 |
| flu | [fluː] | 福禄 | 流感 |
| high | [haɪ] | 嗨 | 高的，高级的 |
| fever | [ˈfiːvə] | 飞舞 | 发烧，发热 |
| headache | [ˈhedeɪk] | 海德克 | 头疼，麻烦 |
| air | [eə] | 爱尔 | 空气，大气 |

生理构造及常见疾病名称

| 英语单词 | 音标 | 发音参考 | 中文翻译 |
|---|---|---|---|
| school | [skuːl] | 斯库尔 | 学校，学院 |
| doctor | [ˈdɒktə] | 多科特 | 医生，博士 |
| where | [weə] | 万阿 | 在哪里 |
| find | [faɪnd] | 范德 | 查到，找到，发现 |
| anybody | [ˈenɪbɒdi] | 爱宁保德 | 任何人 |
| stay | [steɪ] | 思德 | 停留，坚持 |
| eat | [iːt] | 伊特 | 吃，进食 |
| smoking | [ˈsməʊkɪŋ] | 思莫金 | 抽烟，冒烟 |
| enough | [ɪˈnʌf] | 伊那夫 | 足够地，充足地 |
| these | [ðiːz] | 西子 | 这些，这些的 |
| over | [ˈəʊvə] | 欧无 | 结束，越过 |

## Useful Sentences 实用例句

If you are sick, you'd better see our school doctor.
如果你们谁生病了，最好去看我们的校医。
Let's learn about our body.　让我们了解一下我们的身体。

## Grammar 语法

**副词**

疑问副词：用来引导特殊疑问句的副词，有 how，where，when，why。

例句：
How are you doing lately?　你最近怎么样？
Where is your son?　你儿子在哪里？
When will you come tomorrow?　你明天什么时候来？
Why are you late today?　你今天为什么迟到？

连接副词：连接副词和疑问副词是相同的词，但连接副词用于引导从句或者与不定式连用。

例句：
Tell me when you will come.  告诉我你什么时候会来。
I don't know where he lives.  我不知道他住在哪里。
That's why I came today.  这就是为什么我今天来。

关系副词：关系副词和疑问副词是相同的词，引导从句做宾语。

例句：
This is the school where I studied for 4 years.  这是我上了4年学的学校。
I forgot when you told me.  我忘了你什么时候告诉我的。

## Extension  扩展

常见疾病

| 英语单词 | 音标 | 发音参考 | 中文翻译 |
| --- | --- | --- | --- |
| cold | [kəuld] | 扣尔德 | 感冒 |
| vomit | ['vɒmɪt] | 我迷特 | 呕吐 |
| toothache | ['tuːθeɪk] | 吐赛课 | 牙疼 |
| diarrhea | [ˌdaɪə'riə] | 戴尔瑞尔 | 腹泻 |
| cramps | [kræmps] | 克乱普思 | 抽筋 |
| asthma | ['æsmə] | 爱思莫 | 哮喘 |

## Exercises  练习

Dialogue Practice  对话练习

1. Lisa, I have a free health check for you, when are you free?
2. Hi Thomas, could you please show me your nose and hands?
3. Could you please always wash hands before cooking?
4. Sorry, I have been coughing for a few days.
5. No smoking is allowed in our home.

生理构造及常见疾病名称

**Possible Answers** 参考答案

1. Lisa, I have a free health check for you, when are you free? —Thank you, I am free anytime.

2. Hi Thomas, could you please show me your nose and hands? —OK.

3. Could you please always wash hands before cooking? —Yes, of course.

4. Sorry, I have been coughing for a few days. —You'd better see a doctor.

5. No smoking is allowed in our home. —OK, I never smoke.

# Unit 13
# 第 13 课

请假、致谢、致歉等用语
Ask for Leave, Thanks & Apology

**Dialogue 情景对话**

Lisa: Hi Emily, may I ask for leave for 2 weeks.

莉萨：嗨，艾米莉，我能请假两周吗？

Emily: Two weeks? May I know why?

艾米莉：两周？我能知道为什么吗？

Lisa: My mother had an accident last night. She got bumped by a car. And now she is in hospital.

莉萨：我妈妈昨晚发生了意外。她被车子撞了。现在她在医院。

Emily: I am sorry to hear that. You should go right now.

艾米莉：我很抱歉，你应该马上就走。

Lisa: Thank you for your understanding. I will try to be back in 2 weeks.

莉萨：谢谢你的理解，我尽量两周内回来上班。

Emily: It's fine. If you need more time, just let me know. I am OK.

艾米莉：没关系，如果你还需要更多时间，告诉我，我也可以的。

Lisa: Thank you so much!

莉萨：非常感谢！

Emily: And I have friends working in hospital and law firm. If you need any support, please let me know anytime.

艾米莉：我有朋友在医院和律师事务所工作，如果你需要帮忙，随时告诉我。

Lisa: Thank you, I really appreciate it. I just need to go back home to figure out the current situation.

莉萨：谢谢，我很感激。我先需要回老家去搞清楚现在的状况。

Emily: Sure, you can go now, no need to finish today.

艾米莉：好的，你现在可以走了，今天不用做完了。

Lisa: Really, thank you so much!

莉萨：真的吗，太感谢了！

Emily: It's OK. You are part of our family. We do care about your family, too. Let's stay in touch.

艾米莉：没关系。你是我们家的一分子。我们也在乎你的家人。我们保持联系。

Lisa: I will call you when I am back to hometown. Goodbye!

莉萨：我到老家以后给你打电话。再见！

## 请假、致谢、致歉等用语

Emily：Goodbye!
艾米莉：再见！

**New Words 新单词**

| 英语单词 | 音标 | 发音参考 | 中文翻译 |
| --- | --- | --- | --- |
| ask | [ɑːsk] | 阿斯克 | 问，询问 |
| thanks | [θæŋks] | 三克斯 | 谢谢（只用复数形式） |
| apology | [əˈpɒlədʒi] | 阿坡了极 | 道歉 |
| why | [waɪ] | 外 | 为什么 |
| mother | [ˈmʌðə] | 妈泽 | 母亲 |
| accident | [ˈæksɪdənt] | 爱克思等特 | 事故，意外 |
| night | [naɪt] | 耐特 | 晚上，夜晚 |
| she | [ʃiː] | 细 | 她 |
| got | [gɒt] | 高特 | 得到，明白 |
| bump | [bʌmp] | 帮普 | 碰撞；隆起物 |
| car | [kɑː] | 卡 | 汽车，车厢 |
| hospital | [ˈhɒspɪtəl] | 豪思匹特尔 | 医院 |
| sorry | [ˈsɒri] | 搜瑞 | 对不起，遗憾的 |
| hear | [hɪə] | 黑尔 | 听到，听说，听 |
| understanding | [ʌndəˈstændɪŋ] | 昂德思丹顶 | 谅解，理解 |
| try | [traɪ] | 特里 | 努力，实验 |
| fine | [faɪn] | 范儿 | 好的，优良的 |
| much | [mʌtʃ] | 马去 | 非常，很 |
| law | [lɔː] | 罗 | 法律，法规 |
| firm | [fɜːm] | 份儿木 | 公司 |
| support | [səˈpɔːt] | 色坡特 | 支持，支撑 |
| appreciate | [əˈpriːʃieɪt] | 阿普瑞雪尔特 | 欣赏，感激 |
| figure out | [ˈfɪgə] [aʊt] | 菲格尔奥特 | 弄懂，搞清楚 |
| current | [ˈkʌrənt] | 卡润特 | 现在的，通用的 |
| situation | [sɪtjuˈeɪʃn] | 思秋俄迅 | 情况，形式 |
| finish | [ˈfɪnɪʃ] | 菲倪许 | 完成，终止 |

## Useful Sentences 实用例句

I am sorry to hear that.　我很抱歉听到这件事。
I really appreciate it.　我很感激。

## Grammar 语法

**致谢的表达方法**：用于对人表示感谢的词语可以比较简短，也可以比较强烈。
简短的感谢语。
Thanks.
Thank you.

较强烈的感谢方式。
Thank you very much! 非常感谢！
Thanks a lot! 太感谢了！（非正式）
Oh, great! 太棒了！　（非正式）
Many thanks! 很感谢！（较正式）

表示不客气，可以有如下回答。
You are welcome. 别客气。（比较美式）
That's all right. 不用谢。
That's OK. 没什么。
No problem. 没问题。（比较美式，非正式）
Not at all. 用不着客气。（正式）

## Extension 扩展

**英语日期的写法**：英语日期一般是月份加序数词来表示具体的日期。
例子：
1st May　5月1日
3rd August　8月3日

5th October　10月5日
20th December　11月20日

Dialogue Practice　对话练习

1. May I have some water?
2. My little dog got bumped by a car.
3. I am leaving tomorrow, let's stay in touch.
4. When you need help, please tell me.
5. Would you mind if I ask for leave today?

 Possible Answers　参考答案

1. May I have some water?　— OK, help yourself.
2. My little dog got bumped by a car.　— I am sorry to hear this.
3. I am leaving tomorrow, let's stay in touch.　— OK, see you later.
4. When you need help, please tell me.　— Thank you very much.
5. Would you mind if I ask for leave today?　— OK.

# Unit 14
# 第 14 课

旅游度假常用语
Travel & Holidays

### Dialogue 情景对话

Lisa: Good morning, Emily.

莉萨：早上好，艾米莉。

Emily: Good morning, Lisa! We will leave for summer holiday from 4th July.

艾米莉：早上好，莉萨，我们7月4号要去过暑假。

Lisa: OK. Are you going back to the US?

莉萨：好的，你们是回美国吗？

Emily: Yes, for 1 month. Then we will go to France and Germany to visit some friends, about 2 weeks.

艾米莉：是的，回去一个月。然后我们去法国和德国看朋友，大约两周。

Lisa: Oh, great. France is a romantic country.

莉萨：哦，太棒了。法国是个浪漫的国家。

Emily: Yes, we all like France, especially French food. Germany is also good. We may also consider to visit Britain, Sweden, Italy and Spain.

艾米莉：是的，我们都喜欢法国，特别是法国菜。德国也很好，我们可能也会去英国、瑞典、意大利和西班牙。

Lisa: That's a lot. Two weeks may not be enough.

莉萨：好多国家啊，两周可能不够吧。

Emily: Well, Europe is not as big as you think. China is so big, some provinces are even bigger than one country in Europe.

艾米莉：嗯，欧洲没有你想得那么大。中国很大，有些省份比欧洲一个国家都大。

Lisa: Really, I need to study the map better.

莉萨：真的啊，看来我要好好看看地图了。

Emily: And we plan to go to Australia or New Zealand during Christmas, my hometown New York is so cold in winter.

艾米莉：我们圣诞节打算去澳大利亚或者新西兰，我老家纽约冬天太冷了。

Lisa: What a lovely life. I will miss you a lot.

莉萨：生活真是丰富多彩啊。我会想念你们的。

Emily: Me too. I will send you post cards.

艾米莉：我也是。我会给你寄明信片。

Lisa：Thank you.

莉萨：谢谢！

## New Words 新单词

| 英语单词 | 音标 | 发音参考 | 中文翻译 |
| --- | --- | --- | --- |
| travel | [ˈtrævəl] | 催我 | 旅行，游历 |
| holiday | [ˈhɒlɪdeɪ] | 好乐迪 | 假日，节日，休息日 |
| June | [dʒuːn] | 琼 | 六月 |
| France | [frɑːns] | 法兰西 | 法国 |
| Germany | [ˈdʒɜːməni] | 卓木拟 | 德国 |
| visit | [ˈvɪzɪt] | 薇姿特 | 访问，参观 |
| romantic | [rəʊˈmæntɪk] | 罗曼蒂克 | 浪漫的，多情的 |
| country | [ˈkʌntri] | 康翠儿 | 国家，国土 |
| consider | [kənˈsɪdə] | 肯思德 | 考虑，认为 |
| Britain | [ˈbrɪtən] | 不瑞腾 | 英国 |
| Sweden | [ˈswiːdən] | 思位登 | 瑞典 |
| Italy | [ˈɪtəli] | 意特利 | 意大利 |
| Spain | [speɪn] | 思倍恩 | 西班牙 |
| big | [bɪɡ] | 比格 | 大的，重要的 |
| think | [θɪŋk] | 新课 | 认为，想 |
| province | [ˈprɒvɪns] | 普罗万思 | 省份，领域 |
| even | [ˈiːvən] | 伊文 | 甚至，即使 |
| than | [ðæn] | 坦 | 比，超过 |
| Europe | [ˈjʊərəp] | 优尔瑞普 | 欧洲 |
| study | [ˈstʌdi] | 思达迪 | 学习，研究，书房 |
| map | [mæp] | 麦普 | 地图，示意图 |
| Australia | [ɒˈstreɪlɪə] | 澳思翠利亚 | 澳大利亚 |
| New Zealand | [njuː] [ˈziːlənd] | 纽西兰德 | 新西兰 |
| New York | [njuː] [jɔːk] | 纽约克 | 纽约 |
| cold | [kəʊld] | 扣尔德 | 寒冷的，冷淡的 |
| lovely | [ˈlʌvli] | 拉五丽 | 可爱的，令人愉快的 |

家政实用英语

| 英语单词 | 音标 | 发音参考 | 中文翻译 |
| --- | --- | --- | --- |
| life | [laɪf] | 来福 | 生活，生存 |
| post | [pəʊst] | 剖思特 | 邮件；张贴 |

## Useful Sentences 实用例句

What a lovely life.　生活真是丰富多彩啊。

## Grammar 语法

一般用来表示国家名字的词都是名词，如果变形为形容词，则可以表示该国家的，或者该国家的人。变化规律有如下几种：

1. +-ese，例如：China-Chinese，Japan-Japanese，Portugal-Portuguese。

2. +-ish，一般国家名以 land 结尾的，将 land 改为 ish，例如：England-English（保留 l），Finland-Finnish（双写 n）。另有一些特殊情况也将词尾改为 ish，例如：Sweden-Swedish，Spain-Spanish。

3. +-n，一般国家名以 a 结尾的加 n，例如：Russia-Russian，Korea-Korean。

4. +-ian，一般国家名以 y 结尾的变 y 为 ian，例如：Italy-Italian。

还有很多没规律的，例如：Germany-German。

列举一些常见的变化：

| | 国家 | 国籍 |
| --- | --- | --- |
| 中国 | China | Chinese |
| 日本 | Japan | Japanese |
| 美国 | America | American |
| 瑞士 | Sweden | Swedish |
| 法国 | France | French |
| 德国 | Germany | German |
| 英国 | Britain | British |
| 瑞典 | Sweden | Swedish |
| 意大利 | Italy | Italian |
| 西班牙 | Spain | Spanish |
| 新西兰 | New Zealand | New Zealander |

|  | 国家 |  | 国籍 |
|---|---|---|---|
| 澳大利亚 | Australia |  | Australian |
| 俄罗斯 | Russia |  | Russian |
| 印度 | India |  | Indian |
| 加拿大 | Canada |  | Canadian |

| 英语单词 | 音标 | 发音参考 | 中文翻译 |
|---|---|---|---|
| Holland | [ˈhɒlənd] | 后兰的 | 荷兰 |
| Poland | [ˈpəulənd] | 剖兰德 | 波兰 |
| Norway | [ˈnɔːweɪ] | 挪威 | 挪威 |
| Finland | [ˈfinlənd] | 芬兰德 | 芬兰 |
| Russia | [ˈrʌʃə] | 日阿夏 | 俄罗斯 |

Dialogue Practice　对话练习

1. Summer is hot, so we have 2 months holiday.

2. Holland is famous for it's flowers.

3. I will miss you if you go travelling.

4. I will send you post cards when I travel.

5. Would you like to go to America with us during summer holiday?

Possible Answers　参考答案

1. Summer is hot, so we have 2 months holiday.　　— Great, I wish I also have long holiday.

2. Holland is famous for it's flowers.　　— Yes, especially tulips.

3. I will miss you if you go travelling.　　— Me too.

4. I will send you post cards when I travel.　　— Thank you.

5. Would you like to go to America with us during summer holiday?　　— Thank you, I would love to.

# Unit 15
# 第 15 课

工作时间及薪资
Working Hours & Salary

**Dialogue** 情景对话

Emily: Hi Cindy, I have a friend looking for an ayi. Could you please give me a rough idea about ayis' experience and salary expectations?

艾米莉: 嗨,辛迪,我有个朋友在找阿姨,你能大概介绍一下阿姨的经验和工资水平吗?

Cindy: Hi Emily, good to hear from you. Normally it is about RMB 5000-7000 per month for non-English speakers, and RMB 6500-10000 per month for English speakers.

辛迪: 嗨,艾米莉,很高兴你联系我。一般不会说英语的阿姨,工资是5000~7000元/月,会说英语的阿姨工资是6500~10000元/月。

Emily: OK. That's very clear. Is there any difference between house maid and nanny?

艾米莉: 好的,我清楚了。家政阿姨和保姆有区别吗?

Cindy: Yes, there are. House maids usually are good at house work, only help with children sometimes. For nannies, they took professional training for taking care of babies from new-born, and have higher education background.

辛迪: 是的,有的。阿姨擅长做家务,只是有时候帮助照看孩子。保姆专业照料新生儿和大孩子,教育背景也更高一点。

Emily: OK. What is normal working hours?

艾米莉: 好的。一般工作时间是怎样的?

Cindy: Normally 8-10 hours per day, 5-6 days per week are regular.

辛迪: 一般8~10小时一天,一周工作5~6天。

Emily: What if they need overtime?

艾米莉: 如果需要加班呢?

Cindy: All ayis are happy to work overtime if they get paid extra.

辛迪: 所有的阿姨都会愿意加班的,只要支付加班费。

Emily: Got it. What about salary raise?

艾米莉: 了解了。那涨工资呢?

Cindy: Usually ayis expect salary raise once a year, 5%-10% is proper based on ayi's performance.

辛迪: 一般阿姨希望一年涨一次工资,根据工作表现,一般每年涨5%~10%。

Emily: Great. Will ask my friend to contact you.

艾米莉：太棒了。会让我朋友联系你的。

Cindy：Thank you！

辛迪：谢谢！

 New Words 新单词

| 英语单词 | 音标 | 发音参考 | 中文翻译 |
|---|---|---|---|
| give | [gɪv] | 给五 | 给，让步 |
| rough | [rʌf] | 拉夫 | 大致的，粗糙的 |
| idea | [aɪˈdɪə] | 爱蝶儿 | 想法，主意 |
| expectation | [ekspekˈteɪʃn] | 伊克斯派克忒迅 | 期望，预期 |
| normally | [ˈnɔːməli] | 闹莫丽 | 正常地，通常地 |
| non | [nɒn] | 囊 | 不是，非 |
| English | [ˈɪŋglɪʃ] | 英格利息 | 英语，英国人；英国的 |
| clear | [klɪə] | 克莱尔 | 清楚的，清晰的 |
| difference | [ˈdɪfərəns] | 迪夫润思 | 差异，不同，争执 |
| between | [bɪˈtwiːn] | 比囲恩 | 在中间，在……之间 |
| maid | [meɪd] | 梅德 | 女仆，侍女 |
| nanny | [ˈnæni] | 耐尼 | 保姆 |
| sometimes | [ˈsʌmtaɪmz] | 三母太母姿 | 有时候 |
| professional | [prəˈfeʃənəl] | 普瑞范迅诺 | 专业的，职业的 |
| training | [ˈtreɪnɪŋ] | 翠尔宁 | 训练，培养 |
| born | [bɔːn] | 波恩 | 出世；天生的 |
| background | [ˈbækgraʊnd] | 拜克格朗德 | 背景 |
| normal | [ˈnɔːməl] | 诺莫尔 | 正常的，正规的 |
| hour | [ˈaʊə] | 奥尔 | 小时，钟头 |
| regular | [ˈregjʊlə] | 瑞久乐 | 定期的，有规律的 |
| overtime | [ˈəʊvətaɪm] | 欧五太母 | 加班时间，延时 |
| get | [get] | 该特 | 得到，获得 |
| extra | [ˈekstrə] | 埃克斯特 | 特别的，额外的 |
| raise | [reɪz] | 瑞兹 | 提高，筹集 |
| expect | [ɪkˈspekt] | 伊克斯派克特 | 期望，指望 |

## 家政实用英语

| 英语单词 | 音标 | 发音参考 | 中文翻译 |
|---|---|---|---|
| base | [beɪs] | 倍思 | 以……为基础；基础，底部 |
| performance | [pəˈfɔːməns] | 普佛门思 | 绩效，性能，表现 |
| contact | [ˈkɒntækt] | 康泰克特 | 接触，联系 |

### Useful Sentences 实用例句

All ayis are happy to work overtime if they get paid extra.
所有的阿姨都会愿意加班的，只要支付加班费。
What about salary raise?　那涨工资呢？

### Grammar 语法

**分数词**：由基数词和序数词构成，基数词代表分子，序数词代表分母。除了分子为1的情况外，序数词都要用复数形式。

$\dfrac{1}{4}$　one-fourth　　　$\dfrac{5}{9}$　five-ninths

$\dfrac{6}{12}$　six-twelfths　　$3\dfrac{1}{6}$　three and one sixth

**小数的读法和用法**：英语里小数点是 point，如 6.3 的英语表达是 six point three，0.5 的英语表达是 zero point five。

**百分数的用法**：百分数由 percent 来表示，比如百分之八为 eight percent，百分之五十五为 fifty-five percent。

### Extension 扩展

各国货币的英文缩写

| 国家和地区 | 中文名称 | 英文缩写 |
|---|---|---|
| 中国 | 人民币 | CNY |
| 中国香港 | 港币 | HKD |
| 美国 | 美元 | USD |
| 日本 | 日元 | JPY |

| 英国 | 英镑 | GBP |
| 欧盟 | 欧元 | EUR |
| 加拿大 | 加元 | CAD |
| 新加坡 | 新加坡元 | SGD |

## Exercises 练习

Dialogue Practice　对话练习

1. What's your salary expectation?
2. I need a nanny to take care of my son.
3. How long are you willing to work every day?
4. Can you work overtime this weekend?
5. May I expect a salary raise after 1 year?

## Possible Answers 参考答案

1. What's your salary expectation?　　— 6000 per month.
2. I need a nanny to take care of my son.　　— OK, we have good candidates.
3. How long are you willing to work every day?　　— Is 10 hours OK for you?
4. Can you work overtime this weekend?　　— Yes, I am OK.
5. May I expect a salary raise after 1 year?　　— Sure.

# 16

## Unit 16
## 第 16 课

周期性预算及报账用语
Budget & Expense Report

# Dialogue 情景对话

**Emily**: Hi Lisa, do you have 10 minutes to discuss our shopping budget?
**艾米莉**: 嗨,莉萨,你有10分钟聊一下我们的购物预算吗?

**Lisa**: Sure. Here I am.
**莉萨**: 当然有空。我来了。

**Emily**: I noticed the expense last month was 30% more than others'. Can you please explain to me why?
**艾米莉**: 我注意到上个月的支出增长了30%。能请你解释一下为什么吗?

**Lisa**: Oh, yes. The wet market nearby was closed. I can only go shopping at supermarket from last month. The food there are more expensive.
**莉萨**: 当然,家附近的菜市场关闭了。我上个月开始只能去超市买菜。超市的菜贵多了。

**Emily**: OK. Is food price in supermarket 30% higher than that in wet market?
**艾米莉**: 好的。超市的菜要贵30%吗?

**Lisa**: Not that much, 10%-15% more. And the kids playground raised their ticketsprice.
**莉萨**: 没有那么贵,贵10%~15%。孩子们玩的游乐场门票也涨价了。

**Emily**: How much?
**艾米莉**: 多少钱?

**Lisa**: From RMB 30 per time to RMB 50 per time.
**莉萨**: 从30块一次涨到50块一次了。

**Emily**: Oh, that's a lot. Lily and Thomas shall go less in the future.
**艾米莉**: 哦,涨了那么多。莉莉和托马斯以后少去几次。

**Lisa**: And it's summer now. The electricity bill is much higher than that in spring, same for the water bill.
**莉萨**: 而且现在是夏天,电费账单比春天的要高很多,水费也是。

**Emily**: OK. I got it. Do you have all fapiaos (invoices)?
**艾米莉**: 好的。我知道了。你发票都有吗?

**Lisa**: Yes, here you are, all in the note book.
**莉萨**: 当然,发票给你,都在笔记本里面了。

**Emily**: Thank you, Lisa. You are always trustworthy. But we shall figure out how to save money, as everything is getting more and more expensive.

艾米莉：谢谢，莉萨。你一直很诚信可靠。但是我们要想想办法怎么节省开支，因为东西都越来越贵了。

Lisa：I think so, too. Maybe I shall go farther for another wet market.

莉萨：我也这么认为。也许我应该走远点，去找其他菜市场。

Emily：Good idea. You can call our driver to help if it's a bit far.

艾米莉：好主意。菜场如果太远，你可以给司机打电话请他帮忙。

Lisa：Thank you!

莉萨：谢谢！

## New Words 新单词

| 英语单词 | 音标 | 发音参考 | 中文翻译 |
| --- | --- | --- | --- |
| budget | [ˈbʌdʒɪt] | 巴极特 | 预算，安排 |
| expense | [ɪkˈspens] | 伊克思潘思 | 损失，消费，开支 |
| report | [rɪˈpɔːt] | 瑞破特 | 报告，报道 |
| minute | [ˈmɪnɪt] | 米尼特 | 分，分钟 |
| discuss | [dɪˈskʌs] | 迪思卡思 | 讨论，论述 |
| notice | [ˈnəʊtɪs] | 诺提思 | 注意，通知，公告 |
| explain | [ɪkˈspleɪn] | 伊克斯潘雷恩 | 说明，解释 |
| wet | [wet] | 威特 | 潮湿的，有雨的 |
| market | [ˈmɑːkɪt] | 马克特 | 集市，市场，行情 |
| nearby | [ˈnɪəbaɪ] | 内尔白 | 附近的，邻近的 |
| supermarket | [ˈsuːpəmɑːkɪt] | 四有普马克特 | 超级市场，超市 |
| expensive | [ɪkˈspensɪv] | 伊克斯潘西吴 | 昂贵的，费钱的 |
| playground | [ˈpleɪɡraʊnd] | 普雷格让德 | 运动场，操场，游乐场 |
| their | [ðeə] | 再儿 | 他们的，她们的 |
| ticket | [ˈtɪkɪt] | 踢克特 | 票子，入场券，标签 |
| future | [ˈfjuːtʃə] | 富有七 | 将来，未来，前途 |
| electricity | [ˌɪlekˈtrɪsɪti] | 伊莱克吹思提 | 电力，电流 |
| bill | [bɪl] | 比尔 | 账单 |
| spring | [sprɪŋ] | 思普林 | 春季 |

家政实用英语

| 英语单词 | 音标 | 发音参考 | 中文翻译 |
|---|---|---|---|
| invoice | [ˈɪnvɒɪs] | 音沃尔思 | 发票，凭证 |
| book | [bʊk] | 布克 | 书籍，书本 |
| trustworthy | [ˈtrʌstwɜːði] | 川思特翁夕 | 可靠的，可以信赖的 |
| farther | [ˈfɑːðə] | 发则 | 更远的 |
| another | [əˈnʌðə] | 阿那则 | 又一个，另外一个 |
| driver | [ˈdraɪvə] | 拽弗 | 驾驶员，司机 |
| bit | [bɪt] | 比特 | 少量，一点 |
| far | [fɑː] | 伐 | 遥远的，远的 |

## Useful Sentences 实用例句

Here I am. 我来了。

From RMB 30 per time to RMB 50 per time. 从30块一次涨到50块一次了。

Here you are. 给你。

## Grammar 语法

**倒装语序**：英语的基本句型是主语+谓语，称为自然语序。如果谓语提到前面，则为倒装语序。倒装语序一般由 there，here，now，then，so，neither，nor 等引导。

> 例句：
> Here you are. 给你。
> There's the bell. 铃响了。
> Here is the address. 这是地址。
> Now comes your turn. 现在轮到你了。
> I don't like cold weather. ——Neither do I. 我不喜欢寒冷的天气。——我也不喜欢。

## Extension 扩展

价格从低到高的说法

free 免费的

cheap    便宜的

reasonable    合理的

quite expensive    有点儿贵

very expensive    很贵

incredibly expensive    超级贵

Exercises  练习

Dialogue Practice 对话练习

1. Lisa, could you please make a shopping list for me?

2. Could you please tell me how much does it cost for food shopping?

3. I am short of money for shopping, may I have 1000 more?

4. Could you please help me to make a chart for all expenses?

5. I need your help to cut the cost for living.

Possible Answers  参考答案

1. Lisa, could you please make a shopping list for me?    — OK.

2. Could you please tell me how much does it cost for food shopping?    — About 3000 per month.

3. I am short of money for shopping, may I have 1000 more?    — OK, here you are.

4. Could you please help me to make a chart for all expenses?    — OK, will make it in 1 week.

5. I need your help to cut the cost for living.    — OK, I will try my best.

Unit 17
第 17 课

常用急救知识及意外防范用语
First-Aid & Accident Prevention

### Dialogue 情景对话

Lisa: Hi Emily, I had some accidents today, cut my finger, burnt my hand, twisted my ankle and hit my head on the door.

莉萨：嗨，艾米莉，我今天发生了几个意外，切到了手指，手被烫了，扭到了关节，头又撞到门上了。

Emily: Oh, sorry to hear this. What a bad day! You should go to hospital immediately. Shall I call an ambulance?

艾米莉：哦，听到这些很抱歉。真是倒霉的一天。你应该马上去医院。我要叫辆救护车吗？

Lisa: No, thank you! I can go to hospital by myself.

莉萨：不用了，谢谢，我可以自己去医院。

Emily: Wait, I will have my driver to take you to hospital. Before he arrives, please wash your hand with cold water, and I will wrap your finger with bandage.

艾米莉：等一下，我让我司机送你去医院。司机到之前，请你先用冷水洗手，用绷带包好手指。

Lisa: Thank you! I will be very careful next time.

莉萨：谢谢！我下次一定小心。

Emily: Sure, but accidents do happen. I will enroll you for a first-aid course when you recover, so you know what to do next time. Hope for the best, but plan for the worst.

艾米莉：当然，但是意外难以避免。等你恢复了，我给你报一个急救课程，下次你就知道怎么处理了。我们总要抱着最好的希望，但是要做好最坏的打算。

Lisa: Thank you, Emily, you are always so kind.

莉萨：谢谢，艾米莉。你总是这么好。

Emily: We are family now. Let's go! The driver has just arrived.

艾米莉：我们是一家人啊。走吧，司机到了。

Lisa: OK.

莉萨：好的。

## 常用急救知识及意外防范用语

### New Words 新单词

| 英语单词 | 音标 | 发音参考 | 中文翻译 |
| --- | --- | --- | --- |
| first-aid | [ˈfɜːstˈeɪd] | 发思特爱德 | 急救的，急救用的 |
| prevision | [prɪˈvɪʒən] | 普瑞万迅 | 预防，预见 |
| cut | [kʌt] | 卡特 | 切开，切割 |
| finger | [ˈfɪŋə] | 飞音哥 | 手指，指针 |
| burnt | [bɜːnt] | 本恩特 | 燃烧（burn 的过去式）；烧伤的 |
| twist | [twɪst] | 特维斯特 | 扭伤，拧，扭曲 |
| ankle | [ˈæŋkəl] | 安可儿 | 踝关节 |
| hit | [hɪt] | 黑特 | 打击，袭击 |
| immediately | [ɪˈmiːdɪətli] | 一米碟儿特雷 | 立即，立刻 |
| ambulance | [ˈæmbjʊləns] | 安不由冷思 | 救护车 |
| by | [baɪ] | 白 | 通过，经过 |
| myself | [maɪˈself] | 麦塞尔夫 | 我自己，亲自 |
| wait | [weɪt] | 威特 | 等候，推迟 |
| drive | [draɪv] | 拽五 | 开车，驾驶 |
| arrive | [əˈraɪv] | 阿瑞吴 | 到达，成功 |
| wrap | [ræp] | 瑞普 | 包装，包起来 |
| bandage | [ˈbændɪdʒ] | 班迪其 | 绷带 |
| careful | [ˈkeəful] | 凯尔福 | 仔细的，小心的 |
| next | [nekst] | 耐克斯特 | 下一个的；下一个 |
| happen | [ˈhæpən] | 哈盆 | 发生，碰巧 |
| enroll | [ɪnˈrəul] | 因肉尔 | 加入，登记，注册 |
| course | [kɔːs] | 考尔思 | 科目，课程 |
| recover | [rɪˈkʌvə] | 瑞卡我 | 恢复，弥补 |
| hope | [həup] | 侯普 | 希望，期望，信心 |
| best | [best] | 百斯特 | 最好的；最好的事物 |
| worst | [wɜːst] | 味儿斯特 | 最坏的；最坏的事情 |

## Useful Sentences　实用例句

What a bad day.　真是倒霉的一天。

Hope for the best, but plan for the worst.

我们总要抱着最好的希望，但是要做好最坏的打算。

## Grammar　语法

**连词**：用于引导从句的词语。连词是一种虚词，可以连接时间、目的、条件、方式、地点等从句。本课中的 so 是用来引导目的从句的连词，除了 so 外，还有 so that，in order that 等。

例句：

I will enroll you for a first-aid course when you recover, so you know what to do next time.

等你恢复了，我给你报一个急救课程，下次你就知道怎么处理了。

Please practice English every day, so your English will be better.

请每天练习英语，这样你的英语就会进步了。

Can you come early morning so that I can have breakfast.

你能早点来吗？这样我就有早饭了。

## Extension　扩展

意外损伤

| 英语单词 | 音标 | 发音参考 | 中文翻译 |
|---|---|---|---|
| bleeding | [ˈbliːdiŋ] | 不里定 | 流血的 |
| bruise | [bruːz] | 不入子 | 擦伤 |
| hurt | [hɜːt] | 贺特 | 损伤 |

## Exercises　练习

Dialogue Practice　对话练习

1. Don't cut your finger when you use a knife.
2. When you have an accident, better go to hospital immediately.
3. All children need to take the first-aid training.
4. Hope for the best, but plan for the worst.
5. Don't burnt your hands when you are cooking.

Possible Answers 参考答案

1. Don't cut your finger when you use a knife.　　— Thank you.
2. When you have an accident, better go to hospital immediately.　　— OK, I will.
3. All children need to take the first-aid training.　　— Yes, they should.
4. Hope for the best, but plan for the worst.　　— That's true.
5. Don't burnt your hands when you are cooking.　　— No, I won't. Thank you.

# Unit 18
# 第 18 课

## 个人着装及礼貌礼仪用语
## Dress Code Courtesy & Etiquette

**Dialogue** 情景对话

Sarah: Hi everyone, good morning!
莎拉: 嗨,大家早上好!

Children: Good morning, Miss!
孩子们: 老师早上好!

Sarah: We are going to learn how to dress properly today.
莎拉: 我们今天学习一下如何适当着装。

Children: Great.
孩子们: 太棒了。

Sarah: At first I have a question—what would you like to say when you need help?
莎拉: 我首先有个问题,当你需要帮助的时候,你说什么?

Jack: Can you please? Could you please? May I ask you to do me a favor?
杰克: 你可以吗?你可不可以?我能请求您帮忙吗?

Sarah: OK. 'Could' sounds better than 'can', and 'may' is also good. And please use 'please' when you ask for help.
莎拉: 好的。"可不可以"比"可以"更合适,"能不能"也不错,当请求帮助的时候,请用"请"。

Monica: OK. Got it. 'Please' is the magic word whenever you ask for help.
莫妮卡: 好的。知道了。请求帮助的时候,"请"是一个有魔力的词。

Sarah: You are right. We are going to talk about the magic word 2, 'thank you'. When do you say 'thank you'?
莎拉: 你说的很对。我们再来谈谈魔力2号词,"谢谢"。你们都是什么时候说"谢谢"的?

Anu: When I need help, or when I get a present, I say 'thank you'.
阿奴: 当我需要帮助,或者当我得到礼物,我都会说"谢谢"。

Sarah: Yes, say thank you for what other people do for you, including your parents, relatives, friends, and teachers.
莎拉: 是的,要对别人给你的帮助说"谢谢",包括对父母、亲戚、朋友、老师。

Xiao Ming: Thank you, teacher.
小明: 谢谢老师。

个人着装及礼貌礼仪用语

Sarah: And I have a good news. We are going to have an annual party next Friday. All are welcome.

莎拉: 我还有一个好消息。我们下周五有个年度聚会。欢迎大家来参加。

Children: Wow, what a great news! What shall we dress?

孩子们: 哇,太棒了!我们该穿什么呢?

Sarah: For boys, please wear suits; for girls, please wear formal dress, no jeans, no sport shoes, OK?

莎拉: 男孩请着西服,女孩请穿正式的裙装,不要穿牛仔,也不要穿运动鞋,好吗?

Children: OK. we will dress formally.

孩子们: 好的,我们都会穿正装。

Sarah: Right, you can dress casual when you come to school, unless it's school day then you have to wear uniform. And formal suits and dresses for formal party, wedding, funeral, or important ceremonies, etc.

莎拉: 对的,你们来学校可以穿休闲装,除非是学校日,只能穿校服。正式的聚会、婚礼、葬礼或者正式的仪式,都要穿正式的西服和裙子。

Lily: My mother always tell me, as students, we are not supposed to dress too sexy or too shallow.

莉莉: 我妈妈一直跟我说,学生不能穿太性感,或者太肤浅。

Sarah: Your mother is correct. Now class is over, see you tomorrow.

莎拉: 你妈妈说得对。现在下课,明天见。

Children: See you tomorrow, Miss.

孩子们: 明天见,老师。

### New Words 新单词

| 英语单词 | 音标 | 发音参考 | 中文翻译 |
| --- | --- | --- | --- |
| dress | [dres] | 德雷思 | 连衣裙,女装 |
| code | [kəʊd] | 扣德 | 代码,密码 |
| courtesy | [ˈkɜːtɪsi] | 可提思 | 礼貌 |
| etiquette | [ˈetɪket] | 爱提凯特 | 礼仪,礼节 |
| properly | [ˈprɒpəli] | 普若普丽 | 适当地,正确地 |
| favor | [ˈfeɪvə] | 飞舞 | 帮助,喜欢 |

107

# 家政实用英语

| 英语单词 | 音标 | 发音参考 | 中文翻译 |
|---|---|---|---|
| magic | [ˈmædʒɪk] | 麦杰克 | 魔术，魔法 |
| people | [ˈpiːpəl] | 皮婆儿 | 人，人类，公民 |
| parent | [ˈpeərənt] | 皮尔润特 | 父母亲 |
| news | [njuːz] | 牛兹 | 新闻，消息 |
| annual | [ˈænjuəl] | 爱牛尔 | 年度的，每年的 |
| Friday | [ˈfraɪdeɪ] | 福来得 | 星期五 |
| boy | [bɒɪ] | 波尔 | 男孩，男人 |
| wear | [weə] | 威尔 | 穿着 |
| girl | [gɜːl] | 够儿 | 女孩，姑娘 |
| formal | [ˈfɔːməl] | 佛墨儿 | 正式的，拘谨的 |
| sports | [spɔːts] | 思波次 | 运动，运动会（sport 的复数）|
| shoes | [ˈʃuːz] | 修字 | 鞋子（shoe 的复数）|
| formally | [ˈfɔːməli] | 佛莫丽 | 正式地，形式地 |
| casual | [ˈkæʒjuəl] | 开秀儿 | 随便的，非正式的 |
| unless | [ʌnˈles] | 昂来思 | 除非，如果不 |
| uniform | [ˈjuːnɪfɔːm] | 尤你孚尔默 | 制服；统一的 |
| wedding | [ˈwedɪŋ] | 外定 | 婚礼，婚宴 |
| funeral | [ˈfjuːnərəl] | 富有那润 | 葬礼，麻烦事 |
| ceremony | [ˈserɪməni] | 塞瑞么尼 | 典礼，仪式 |
| etc | [ɪtˈsetərə] | 伊特塞特若 | 等等，其他 |
| sexy | [ˈseksi] | 塞克锡 | 性感的，迷人的 |
| shallow | [ˈʃæləʊ] | 沙洛 | 肤浅的，浅的 |
| student | [ˈstjuːdənt] | 思丢登特 | 学生，学者 |
| suppose | [səˈpəʊz] | 色坡子 | 假设，猜想，应当 |

 **Useful Sentences** 实用例句

'Could' sounds better than 'can', and 'may' is also good.
"可不可以"比"可以"更合适，"能不能"也不错。
Now class is over.　现在下课。

**Grammar 语法**

引导时间从句的连词有 when, whenever, while, as, before, after, until, since 等。主要表示时间的先后顺序。

> 例句：
> 'Please' is the magic word whenever you ask for help.
> 当你们请求帮助的时候，"请"是一个有魔力的词。
> Look before you leap.
> 三思而后行。
> I will tell you after he leave.
> 他走了我再告诉你。
> It has been ten years since she came to Shanghai.
> 她来上海十年了。

并列连词：主要用来表示并列、转折、因果等关系。常用的有 and, or, but, however 等。

> 例句：
> When I need help, or when I get a present, I always say thank you.
> 当我需要帮助，或者当我得到礼物，我都会说谢谢。
> Lily likes singing and dancing.
> 莉莉喜欢唱歌和跳舞。
> She works very hard, but still can't reach the high standard.
> 她很努力，但仍然不能达到高标准。

 **Extension 扩展**

情态动词 can, could, may, might 的区别

can 和 could 一般表示说话人的许可，could 比 can 更加犹豫，也显得更加婉转。

Can I ask you a question?　我能问你个问题吗？

Can you come tomorrow?　你明天能来吗？

Could you help me?　你能帮我吗?

may 和 might 也是指许可，与 can 和 could 相比，may 和 might 更正式、更尊敬，might 表示的口气比 may 更婉转，可以用来代替 may。

May I come in?　我可以进来吗?

May I make a suggestion?　我可以提个建议吗?

Might I borrow your book?　我可以借你的书吗?

  Exercises 练习

Dialogue Practice　对话练习

1. Hi, could you please do me a favor?

2. I have a good news for you.

3. I love my family, including my parents and brother.

4. Do you have suits for party?

5. When class is over, you can go home.

 Possible Answers 参考答案

1. Hi, could you please do me a favor?　— Yes, of course.

2. I have a good news for you.　— What news?

3. I love my family, including my parents and brother.　— Me too.

4. Do you have suits for party?　— Yes, I have.

5. When class is over, you can go home.　— Thank you.

# 19

**Unit 19**
**第 19 课**

宗教习俗及相关忌讳用语
Religious Custom & Taboo

### Dialogue 情景对话

Sarah: Good morning, everyone!
莎拉: 大家早上好。

Children: Good morning, Miss.
孩子们: 老师早上好。

Sarah: Today we are going to discuss about our religion. Is it OK?
莎拉: 今天我们来讨论一下我们的信仰, 可以吗?

Children: Of course.
孩子们: 当然可以。

Jack: Let me start first. I am Christian. I go to church every Sunday. And we believe in Jesus Christ, who requires all of us to do only good things.
杰克: 我先开始, 我是基督教徒, 我每周日去教堂。我们信仰基督耶稣, 他教导我们只做好事。

Monica: I am Catholic, I also go to Church every Sunday, but a different Church.
莫妮卡: 我是天主教徒, 我也每周日去教堂, 但是是另外一个教堂。

Sarah: Yes, we have different religions, but we shall respect each other's.
莎拉: 是的, 我们有不同的信仰, 但是我们都要尊重彼此的信仰。

Anu: I am a Hindu, and all my family members are vegetarian. We don't eat any meat, no milk or egg.
阿奴: 我是印度教徒, 我们家人都是素食主义者, 不吃任何肉、牛奶或者鸡蛋。

Xiao Ming: I don't have any religion yet, but my grandma believes in Buddhism. She goes to the temple in the middle and end of the month.
小明: 我还没有什么信仰呢, 但是我奶奶信仰佛教, 她每两周去一次寺庙。

Muhammad: I believe in Islam, though some extreme organizations call themselves Islam, we don't recognize them, and we all love peace.
穆罕穆德: 我信仰伊斯兰教, 虽然有些极端组织也自称伊斯兰教徒, 但我们不认可他们, 我们热爱和平。

Sarah: Great, You are all good children. Although we have different religions, we all love peace. We shall respect each other and live friendly together.
莎拉: 太棒了, 你们都是好孩子。虽然我们信仰不同, 但是我们都热爱和平。我们

## 宗教习俗及相关忌讳用语

应当互相尊重，和平相处。

**Children**：Yes, we all love peace.

**孩子们**：对的，我们都热爱和平。

 New Words 新单词

| 英语单词 | 音标 | 发音参考 | 中文翻译 |
| --- | --- | --- | --- |
| religious | [rɪˈlɪdʒəs] | 瑞丽群思 | 宗教的 |
| custom | [ˈkʌstəm] | 卡思特木 | 习惯，惯例 |
| taboo | [təˈbuː] | 特布 | 禁忌 |
| religion | [rɪˈlɪdʒən] | 瑞丽群 | 宗教，信仰 |
| Christian | [ˈkrɪstiən] | 克里斯提安 | 基督徒，信徒 |
| church | [tʃɜːtʃ] | 丘奇 | 教堂，礼拜 |
| Sunday | [ˈsʌndeɪ] | 桑德 | 星期天 |
| believe | [bɪˈliːv] | 比利五 | 信任，相信，认为 |
| Jesus | [ˈdʒiːzəs] | 吉瑟斯 | 耶稣，上帝之子 |
| Christ | [kraɪst] | 克瑞斯特 | 基督，救世主 |
| require | [rɪˈkwaɪə] | 瑞块尔 | 需要，要求，命令 |
| thing | [θɪŋ] | 欣 | 东西，事情 |
| Catholic | [kæθəlɪk] | 凯瑟立克 | 天主教的；天主教徒 |
| different | [ˈdɪfərənt] | 迪福瑞特 | 不同的，特别的 |
| Hindu | [ˈhɪnduː] | 印度 | 印度教，印度人，印度教徒 |
| vegetarian | [ˌvedʒɪˈteəriən] | 万吉泰尔瑞安 | 素食者，食草动物；素食的 |
| egg | [eg] | 艾格 | 鸡蛋 |
| yet | [jet] | 耶特 | 还，但是 |
| grandma | [ˈgrændmaː] | 格兰德玛 | 奶奶，外婆 |
| Buddhism | [ˈbʊdɪz(ə)m] | 布地真 | 佛教 |
| temple | [ˈtempl] | 坦普尔 | 寺庙，庙宇 |
| middle | [ˈmɪdl] | 米德尔 | 中间的，中央的；中央 |
| Islam | [ˈɪzlɑːm] | 伊斯兰 | 伊斯兰教 |

113

## 家政实用英语

| 英语单词 | 音标 | 发音参考 | 中文翻译 |
|---|---|---|---|
| though | [ðəʊ] | 仇 | 虽然，可是 |
| extreme | [ɪkˈstriːm] | 伊克斯坠木 | 极端的，极度的 |
| organization | [ˌɔːɡənaɪˈzeɪʃn] | 奥格耐追寻 | 组织，机构 |
| themselves | [ðəmˈselvz] | 真木赛尔吴兹 | 他们自己，她们亲自 |
| recognize | [ˈrekəɡnaɪz] | 瑞克格耐兹 | 认出，识别 |
| peace | [piːs] | 皮斯 | 和平，安静 |
| although | [ɔːlˈðəʊ] | 奥尔走 | 尽管，虽然 |
| live | [lɪv] | 立吴 | 经历，生活 |
| friendly | [ˈfrendli] | 弗兰德利 | 友好的，亲切的 |

### Useful Sentences 实用例句

Is it OK? 可以吗？

We have different religions, but we shall respect each other's.
我们有不同的信仰，但是我们都要尊重彼此的信仰。

### Grammar 语法

引导让步从句的连词主要有 although, though, even though, even if, while 等，表示"虽然""尽管""即便"等。

---

例句：

I believe in Islam, though some extreme organizations call themselves Islam, we don't recognize them.

我信仰伊斯兰教，虽然有些极端组织也自称伊斯兰教徒，但我们不认可他们。

I will not lend money to you even if you are my friend.

即使你是我朋友，我也不会借钱给你。

Although it is raining outside, he still wants to go out.

虽然外面在下雨，他还是想要出去。

### Exercises 练习

**Dialogue Practice**　对话练习

1. Do you have any religion?
2. We go to Church every Sunday. Do you want to come together?
3. My husband is a vegetarian, please only cook vegetables.
4. There are many beautiful temples in China.
5. We all love peace.

### Possible Answers 参考答案

1. Do you have any religion?　　— No, I don't.
2. We go to Church every Sunday. Do you want to come together?　　— No, thank you.
3. My husband is a vegetarian, please only cook vegetables.　　— OK.
4. There are many beautiful temples in China.　　— Yes.
5. We all love peace.　　— Me too.

# Unit 20
# 第 20 课

请 辞 用 语
Resignation & Dismissal

家政实用英语

**Dialogue 情景对话**

Lisa: Hi, Emily, may I have a talk with you?
莉萨：嗨，艾米莉，我能和你谈一下吗？

Emily: Sure, what's up?
艾米莉：当然，怎么了？

Lisa: Well, my son is going to primary school this September.
莉萨：我儿子9月要去小学了。

Emily: I know. That's good!
艾米莉：我知道。这很好啊！

Lisa: Unfortunately, I can't get him a place in a local primary school. Local schools can only accommodate with local kids. There are not enough places for children from other provinces.
莉萨：不幸的是，本地小学都满了，我儿子进不去。学校只能协调所有本地孩子去上学，没有足够的位子给我们其他省份的孩子。

Emily: Oh, that's really annoying. Maybe your son can go to private school.
艾米莉：哦，这个真不好。或许你儿子可以去私立学校。

Lisa: I wish I can afford. Private schools are too expensive for my family. In this case, I have to go back to our hometown by the end of August. Now I have to resign. I will leave when you find a good replacement ayi.
莉萨：我希望我能付得起，私立学校对我们来说太贵了。这样的话，我只能在8月底回老家去了。现在我只能辞职了，等你找到合适的阿姨我再走。

Emily: Oh, so sorry to hear this. You are my family member now. Lily will cry if she knows this.
艾米莉：哦，听到这个消息我真是抱歉，你是我们的家庭成员了，莉莉如果知道会哭的。

Lisa: I'm very sorry about this, but my son is very important to me, too.
莉萨：我非常抱歉，但是我儿子对我来说也很重要。

Emily: I understand, just feel sorry. I will host a farewell party for you. Please bring your son and your husband then!
艾米莉：我理解，我只是觉得有点遗憾。我会为你开个送别聚会，到时带上你儿子和你

118

老公。

Lisa: Thank you for your invitation. I will miss you a lot.
莉萨：谢谢你的邀请，我会想念你们的。

Emily: Me too, you can come to visit us, we can also go to visit you.
艾米莉：我也是，你可以来看我们，我们也可以去看你。

Lisa: Definitely, I will come to visit you when my son is settled in school.
莉萨：当然了，等我儿子学校安顿好了，我就来看你们。

 New Words 新单词

| 英语单词 | 音标 | 发音参考 | 中文翻译 |
| --- | --- | --- | --- |
| resignation | [ˌrezɪɡˈneɪʃən] | 瑞兹个内迅 | 辞职，放弃 |
| dismissal | [dɪsˈmɪsl] | 迪斯米索尔 | 解雇，辞退 |
| up | [ʌp] | 阿普 | 起来，向上 |
| son | [sʌn] | 桑 | 儿子，孩子 |
| primary | [ˈpraɪməri] | 普瑞么瑞 | 主要的，初级的 |
| September | [sepˈtembə] | 色普填补儿 | 九月 |
| unfortunately | [ʌnˈfɔːtʃənətli] | 昂佛群那特丽 | 不幸地 |
| place | [pleɪs] | 普雷思 | 席位，座位，地方 |
| local | [ˈləʊkəl] | 楼克尔 | 本地的，当地的；当地居民 |
| accommodate | [əˈkɒmədeɪt] | 阿扣莫德特 | 容纳，供应 |
| annoying | [əˈnɔɪɪŋ] | 阿诺儿英 | 讨厌的 |
| private | [ˈpraɪvət] | 普热爱威特 | 私人的，私有的 |
| wish | [wɪʃ] | 维洗 | 希望，祝福 |
| afford | [əˈfɔːd] | 阿佛德 | 负担得起，提供，给予 |
| case | [keɪs] | 凯斯 | 情况，箱子 |
| end | [end] | 安德 | 结束，尽头，目标 |
| August | [ɔːˈɡəst] | 奥格斯特 | 八月 |
| resign | [rɪˈzaɪn] | 瑞赞尔 | 辞职，放弃 |
| replacement | [rɪˈpleɪsmənt] | 瑞普雷思门特 | 更换，替换 |
| member | [ˈmembə] | 馒不 | 成员，会员 |
| important | [ɪmˈpɔːtənt] | 因母坡腾特 | 重要的，重大的 |

## 家政实用英语

| 英语单词 | 音标 | 发音参考 | 中文翻译 |
|---|---|---|---|
| understand | [ˌʌndəˈstænd] | 昂德思丹德 | 理解，获悉 |
| feel | [fiːl] | 菲尔 | 感觉，认为 |
| farewell | [feəˈwel] | 菲尔威尔 | 告别，辞别 |
| bring | [brɪŋ] | 布林 | 带来，促使 |
| invitation | [ˌɪnvɪˈteɪʃən] | 因维特勋 | 邀请，请帖 |
| settle | [ˈsetl] | 塞特尔 | 解决，定居 |

### Useful Sentences 实用例句

what's up?  怎么了？

I wish I can afford.  我希望我能承担得起。

So sorry to hear this.  听到这个消息我真是遗憾。

### Grammar 语法

虚拟语气：表示一种假象的情况或者主观愿望，动词要用一种特殊形式。虚拟语气分为三类。

1. 现在虚拟语气：主要用动词原形。

I wish I can afford.

God bless you.

2. 过去虚拟语气：和陈述语气的过去式相同，但是动词 be 要用 were 的形式。

If only I were not so sad. 我要是不那么悲伤就好了。

Imagine you played football at school. 设想你在学校踢足球。

3. 过去完成形式：表示虚拟的过去已经完成的动作。

I wish I had finished my homework. 真希望我已经做完作业了。

If I had met her, I would have told you. 如果我见到她了，我就会告诉你了。

### Extension 扩展

四季：一年有四个季节，一般春天是 3—5 月，夏天是 6—8 月，秋天是 9—11 月，冬天是 12—2 月。

spring 春天　　summer 夏天
autumn 秋天　　winter 冬天

## Exercises 练习

Dialogue Practice　对话练习

1. You are crying, what's up?
2. Winter is very annoying in Shanghai, too cold.
3. Children are important to each family.
4. Thanks for your invitation, will arrive on time.
5. I will miss you when you are gone.

## Possible Answers 参考答案

1. You are crying, what's up?　　— Sorry, my bike is broken.
2. Winter is very annoying in Shanghai, too cold.　　— I agree.
3. Children are important to each family.　　— Yes, I think so.
4. Thanks for your invitation, will arrive on time.　　— See you then.
5. I will miss you when you are gone.　　— Me too.

# 附 录

## 音 标

音标分为元音（长元音、短元音、双元音）和辅音（浊辅音、清辅音）。

### 长元音

| 音标 | 参考发音 | 发音方式 | 例词 | 读音 |
| --- | --- | --- | --- | --- |
| iː | 亿 | 发音时间略长 | see | [siː] |
| ɑː | 啊 | 嘴巴张大，发音时间长 | arm | [ɑːm] |
| ɔː | 哦 | 声音拖长 | law | [lɔː] |
| uː | 雾 | 声音拖长 | too | [tuː] |
| ɜː | 饿 | 发音时间稍长 | bird | [bɜːd] |

### 短元音

| 音标 | 参考发音 | 发音方式 | 例词 | 读音 |
| --- | --- | --- | --- | --- |
| ɪ | 衣 | 急促地发音 | sit | [sɪt] |
| e | 哎 | 嘴唇扁平地发音 | ten | [ten] |
| æ | 哎 | 口型张大，发音清脆 | hat | [hæt] |
| ɒ | 哦 | 发音急促 | got | [gɒt] |
| ʊ | 乌 | 发音急促 | put | [pʊt] |
| ʌ | 阿 | 嘴微微张开，发音短促 | cup | [kʌp] |
| ə | 额 | 发短音 | ago | [əˈgəʊ] |

# 家政实用英语

## 双元音

| 音标 | 参考发音 | 发音方式 | 例词 | 读音 |
|---|---|---|---|---|
| eɪ | 诶 | 发音短促，响亮 | page | [peɪdʒ] |
| aɪ | 爱 | 拉长发音，逐步收缩 | five | [faɪv] |
| ɔɪ | 哦 | 衣连着读 | boy | [bɔɪ] |
| əʊ | 呕 | 用力发音，清晰 | home | [həʊm] |
| aʊ | 傲 | 发音前轻后重 | now | [naʊ] |
| ɪə | 衣饿 | 连着读，前短后长 | near | [nɪə] |
| eə | 哎尔 | 连着读，前轻后重 | hair | [heə] |
| ʊə | 悟尔 | 连着读，前轻后重 | pure | [pjʊə] |

## 浊辅音

| 音标 | 参考发音 | 发音方式 | 例词 | 读音 |
|---|---|---|---|---|
| b | 不 | 轻而快发音 | bad | [bæd] |
| d | 得 | 声音短促发音 | dad | [dæd] |
| g | 哥 | 声音短促，声带震动 | girl | [gɜːl] |
| v | 五 | 上牙压着下唇发音 | very | ['veri] |
| ð | 斯 | 上齿咬舌头 | this | [ðɪs] |
| z | 滋 | 舌头不碰上下口腔，声带震动 | zoo | [zuː] |
| r | 日 | 舌头两边卷起来 | red | [red] |
| dʒ | 姬 | 清晰响亮，声带震动，但不送气 | job | [dʒɒb] |
| j | 叶 | 轻轻发音，舌头下沉 | year | [jɪə] |
| w | 呜 | 发音短而快 | what | [wɒt] |
| m | 木 | 嘴巴闭住，气流从鼻子出来 | mother | ['mʌðə] |
| n | 呢 | 嘴巴微张，舌尖顶住上颚，气流从鼻子出来 | new | [njuː] |

| 音标 | 参考发音 | 发音方式 | 例词 | 读音 |
|---|---|---|---|---|
| ŋ | 昂 | 嘴巴张大，舌头向下弯曲，气流从鼻子出来 | long | [lɒŋ] |
| l | 了 | 将舌头抵住上齿龈上，再向下轻拨，震动声带 | leave | [liːv] |

### 清辅音

| 音标 | 参考发音 | 发音方式 | 例词 | 读音 |
|---|---|---|---|---|
| p | 普 | 轻轻地发音 | pen | [pen] |
| t | 特 | 轻轻地发音 | tea | [tiː] |
| k | 柯 | 轻轻地发音 | kiss | [kɪs] |
| f | 福 | 声音短促，不震动声带 | food | [fuːd] |
| θ | 思 | 上齿咬舌头，轻轻发音 | thank | [θæŋk] |
| s | 丝 | 轻轻发音，不震动声带 | sing | [sɪŋ] |
| ʃ | 施 | 轻轻发音 | she | [ʃiː] |
| ʒ | 衣 | 舌头不要碰到上下左右口腔 | pleasure | [ˈpleʒə] |
| h | 喝 | 轻轻发音 | home | [həʊm] |
| tʃ | 七 | 清晰响亮 | cheap | [tʃiːp] |

### 26个字母音标

A a [eɪ]　　B b [biː]　　C c [siː]　　D d [diː]　　E e [iː]
F f [ef]　　G g [dʒiː]　　H h [eɪtʃ]　　I i [aɪ]　　J j [dʒeɪ]
K k [keɪ]　　L l [el]　　M m [em]　　N n [en]　　O o [əʊ]
P p [piː]　　Q q [kjuː]　　R r [ɑː]　　S s [es]　　T t [tiː]
U u [juː]　　V v [viː]　　W w [ˈdʌbljuː]　　X x [eks]　　Y y [waɪ]
Z z [ziː] [zed]